Signs of the Zodiac:
floor mosaic, Beth-Aleph Synagogue, Palestine (5th century)

One of the first forms of calendar, symbolic here of the struggles of
the early mathematicians to impose a pattern and an order on nature

Games, Gods and Gambling

A History of Probability and Statistical Ideas

F. N. David

DOVER PUBLICATIONS, INC.
Mineola, New York

Published in Canada by General Publishing Company, Ltd., 30 Lesmill Road, Don Mills, Toronto, Ontario.

Published in the United Kingdom by Constable and Company, Ltd., 3 The Lanchesters, 162–164 Fulham Palace Road, London W6 9ER.

Bibliographical Note

This Dover edition, first published in 1998, is an unabridged republication of the work first published by Charles Griffin & Co. Ltd., London, in 1962.

Library of Congress Cataloging-in-Publication Data

David, F. N. (Florence Nightingale), 1909–
    Games, gods, and gambling : a history of probability and statistical ideas / F. N. David.
        p. cm.
    "An unabridged republication of the work first published by Charles Griffin & Co. Ltd., London, in 1962"—T.p. verso.
    Includes bibliographical references (p.   –   ) and index.
    ISBN 0-486-40023-9 (pbk.)
    1. Probabilities—History. 2. Games   of   chance (Mathematics)—History. 3. Statistics—History. I. Title.
QA273.A4D374   1998
519.2—dc21   97-46416
                                                                         CIP

Manufactured in the United States of America
Dover Publications, Inc., 31 East 2nd Street, Mineola, N.Y. 11501

TO

M. G. KENDALL

BECAUSE OF HIS CONTINUING INTEREST AND ENCOURAGEMENT

What shall we tell you? Tales, marvellous tales
Of ships and stars and isles where good men rest,
Where nevermore the rose of sunset pales,
And winds and shadows fall toward the West.

.   .   .   .   .

And how beguile you? Death hath no repose
Warmer and deeper than that Orient sand
Which hides the beauty and bright faith of those
Who made the Golden Journey to Samarkand.

And now they wait and whiten peaceably,
Those conquerors, those poets, those so fair;
They know time comes, not only you and I,
But the whole world shall whiten, here or there:

When those long caravans that cross the plain
With dauntless feet and sound of silver bells
Put forth no more for glory or for gain,
Take no more solace from the palm-girt wells;

When the great markets by the sea shut fast
All that calm Sunday that goes on and on,
When even lovers find their peace at last,
And Earth is but a star, that once had shone.

Prologue: *The Golden Journey to Samarkand*
JAMES ELROY FLECKER (1884–1915)

Preface

TODHUNTER on the "History of Probability" has been and always will be the major work of reference in this subject. He is rarely wrong, always explicit, and has obviously read widely through all available literature. That he stops early in time (with Laplace) and thus does not treat the interesting developments of the nineteenth century is hardly his fault, since he wrote his book in 1864–65 and would have been collecting his material for some years before this. From the point of view of the development of ideas he does not start soon enough. He notes the mathematical arguments fairly and with precision, but this is like embarking on a river when it has become of respectable size, and paying no attention to the multitude of small streams and tributaries of which it is the united outcome.

The idea that one might speculate about the development of the random element through references in literature of all kinds—classical, archaeological, biographical, poetical and fictional—is one which came to me as a student some thirty years ago. Since then, as friends and colleagues learnt of my interest, the number of references to which my attention has been drawn has increased rapidly, each year filling many notebooks. It is too much to hope that this list of references can ever be completed, but it is, I think, fair to say that many of the most significant have been made known to me. And from these references I have tried to supplement Todhunter on the early development of ideas about chance and to fill in a certain amount of the background of ideas and controversies which attended the creation of the mathematical theory of probability.

Writers on the history of science, and in particular on the history of a branch of mathematics, tend to fall into one of two categories : they either write severely on the subject-material and leave the men who created it as wooden puppets, or they tell us interesting (and often apocryphal) stories about personalities, with no real attempt to place their achievements. Thus Todhunter will rarely be read for pleasure, although always for profit, by anyone interested in probability theory, while the potted biographies, always read for pleasure, convey little to us which profits our understanding of theory. I have tried, probably without success, to steer a middle way between the Scylla of Todhunter and the Charybdis of the story-teller. The man creates the mathematical theorem, but the events of a man's life create the man, and the three are indissoluble. Thus I would hold, for example, that it is not enough to remark on and wonder at de Moivre's analytic genius, but that one should also realise that poverty was his spur, that much of his work might not have been achieved had he been sure of a post which would have brought him some leisure and about which he wrote so longingly to John Bernoulli.

Wherever possible I have gone back to the original documents to check mathematical developments; in the same way, I have tried to check the stories told about great men. The similarity between so many of these latter induces a profound scepticism about all of them, even when told autobiographically, and I have tried to indicate this scepticism in the text. I have tried, not invariably with success, never to express an opinion without at least adequate information of the relative facts. The documentation of one's opinions naturally grows easier as the millennium advances, although the upheaval of the French Revolution led to much of value being mislaid. The lack of information about de Moivre can only be described as startling.

When I began work on the history of probability and statistics I had the ambitious aim of covering the whole field from pre-history to the present day and including appraisals of the life and works of such giants as Gauss and Poisson. The accumulation of facts is, however, a slow business, particularly when it is pursued solely as a hobby, and I think I will not achieve my original goal. I have in mind to write a monograph on the great triumvirate, James Bernoulli–Montmort–de Moivre, and another on Laplace. An appreciation of the works of Laplace from the modern statistical aspect is long overdue and would lead to the realisation of how many theorems are still being published which are special cases of his general results. At the rate at which I work these two projects will take me beyond my allotted span of years. There is much still to be done. Not only should the history after Laplace be written, but the time is ripe for a monograph on inverse probability. It is to be hoped that someone may be stimulated to compile this.

So many persons have given me references and have borne with what must have seemed my interminable discussions that space would not permit me to thank them individually. I count myself fortunate in having so many friends. I feel I should add, however, that if it had not been for the continuing interest (and lately the constant prodding) of M. G. Kendall I doubt whether I should ever have stopped collecting information and taken the irretrievable step of putting it on paper; once something is written it dies a little and one is always reluctant to call a halt. I owe him much both in ideas and stimulus, and it is to him therefore that I would dedicate this book.

F. N. DAVID

London,
1961

Contents

Chapter		Page
1	The development of the random event	1
2	Divination	13
3	The probable	21
4	Early beginnings	27
5	Tartaglia and Cardano	40
6	Cardano and *Liber de Ludo Aleae*	55
7	Galileo	61
8	Fermat and Pascal	70
9	The arithmetic triangle and correspondence between Fermat and Pascal	81
10	Bills of Mortality	98
11	Christianus Huygens	110
12	Wallis, Newton and Pepys	123
13	James Bernoulli and *Ars Conjectandi*	130
14	Pierre-Rémond de Montmort and *The Essai d'Analyse*	140
15	Abraham de Moivre and *The Doctrine of Chances* ..	161

CONTENTS

Appendix   *Page*

1 Buckley's *Memorable Arithmetic*   179
   Translated by Jean Edmiston

2 Galileo's *Sopra le Scoperte dei Dadi*   192
   Translated by E. H. Thorne

3 Brother Hilarion de Coste's *Life of Father Marin*
        Mersenne   196
   Translated by Maxine Merrington

4 Letters between Fermat and Pascal and Carcavi   ..   229
   Translated from Oeuvres de Fermat by Maxine Merrington

5 From *The Doctrine of Chances* by A. de Moivre   ..   254

Index   269

List of Plates

Frontispiece   Signs of the Zodiac : floor mosaic, Beth-
Aleph Synagogue, Palestine (5th cen-
tury)

Plate

1   Egyptian tomb-painting showing a
nobleman in after-life using an astragalus
in a board game   ..   ..   ..   *Facing page* 8

2   The board game of " Hounds and
Jackals " (*c.* 1800 B.C.), from the tomb
of Reny-Soube at Thebes, Upper Egypt   *Between pages* 8 *and* 9

3   Four views of an astragalus   ..   ..   *Between pages* 8 *and* 9

4   Die roughly made on classical pattern
Imitation astragalus carved in stone
Egyptian carved ceremonial die, with
typical astragali   ..   ..   ..   *Facing page* 9

5   Lorenzo Leombruno's Allegory of Fortune (16th century)   ..   ..   ..   *Facing page* 24

6 and 7   Excerpt from the manuscript poem
" De Vetula " (Harleian MS. 5263) giving the number of ways in which three
dice can fall   ..   .. *Between pages* 24 *and* 25

8   Fra Luca Paccioli (National Galleries
of Capodimonte, Naples)   ..   ..   *Facing page* 25

9   Abraham de Moivre—painted by Jos.
Highmore, 1736   ..   ..   ..   *Facing page* 160

Acknowledgements

ACKNOWLEDGEMENTS of permission to use quoted matter have in most cases been given in footnotes. Thanks are also due to the following persons not otherwise mentioned: Messrs Martin Secker & Warburg Ltd. for the introductory quotation from J. E. Flecker; the Secretary of the Netherlands Society for the Sciences for the use of extracts from the commentaries on the writings of Christianus Huygens; the Editor of *Biometrika* for the use of material given in Chapters 1, 2 and 3; the Curator of Greek and Roman Antiquities of the British Museum; and the Yates Professor of Classical Archaeology, University College, London, for permission to photograph the Egyptian die and astragali shown in Plate 4. Apologies are offered in advance if, by inadvertence, any quotation from copyright matter has been unacknowledged.

Miss J. Townend made the line drawings for Chapter 1. Grateful thanks are due to the translators of the Appendices— two of whom have also assisted with the indexing. Finally I am indebted to E. V. Burke for his patience, attention to detail, and unfailing help throughout both the printing and the preparation of the manuscript.

chapter **1**

The development of the random event

> The thing which hath been, it is that which shall be, and
> that which is done is that which shall be done ; and there
> is no new thing under the sun.
>
> ECCLESIASTES, i. 9.

It is interesting to speculate on the development of counting and
the final adoption of the decimal notation for the writing down
of numbers. Historians do not appear to have reached any firm
conclusion about this, although counting in a scale of ten is
universal among civilised people and nearly universal among
primitive tribes. The suggestion that the decimal scale arose from
the accident that we possess four fingers and one thumb on each
hand is one which has a certain credibility.* The ancient Persians
and Greeks, to quote two examples, had words for five which
meant " hand ", and the Roman symbol V for five is supposed—
on what authority one is not sure—to represent the V between
the thumb and the forefinger. There are many references to
counting in ancient literature all of which refer to the decimal
scale. The best known of these perhaps is that given in some
translations of the *Odyssey* in which Proteus is depicted as counting
his seals by fives.

The one-to-one correspondence between the fingers of the
hand and the objects to be enumerated may have taken hundreds
or thousands of years to establish itself, but an even greater

---

* I would not put it higher than this. To me it would appear more " natural "
if we counted in a scale of 4, from the four fingers, with the thumb indicating
that the set of 4 is complete.

1

conceptual difficulty was possibly met in the extension of this idea and in the representation of a number. The several joints of the fingers were pressed into notational use, and for at least some time the hand placed in definite positions on the human body stood for various multiples of fives and tens. The Venerable Bede, writing in the eighth century, could count up to a million in this way. The present-day system of writing down numbers is thought to have originated in India (or possibly Tibet) in the ninth or tenth centuries. It was slow to spread into general use and obviously caused great difficulty, as may be seen from William Buckley, writing a little before 1567.* The comparatively little arithmetic progress made by the Greeks and Romans would almost undoubtedly have been because of their cumbersome notation. The marvellous Greek mathematicians of the years 500–400 B.C. carried out their calculations with coloured stones, forerunners of the abacus which is still in use in the East today.

It is a commonplace for archaeologists to report from excavations that small quantities of stones of different colours—obviously collections of some kind—have been found. These stones may have been the tally-stones used in some kind of primitive counting—they may even have been money—or they may have been counters used in an early game of chance. It is also a commonplace for finds of large numbers of the astragalus or heel-bone,† to be reported, particularly those of animals with hooves such as all varieties of deer. The astragali are reported to be much more numerous than other kinds of bone. This in itself is not surprising. The astragalus is solid and would not be coveted for the sake of its bone-marrow content ; it is in fact very difficult to smash, even deliberately.

---

* Appendix 1.

† The words astragalus, talus, huckle-bone and knuckle-bone appear to be used indiscriminately in both ancient and medieval literature. The astragalus is a bone in the heel lying above the talus, which latter, in strict anatomical sense, is the heel-bone. The huckle-bone is the astragalus, while the knucklebone, strictly the bone in the hand, has been used to mean huckle-bone also since the sixteenth century. In the references in the literature to gaming, and indeed possibly in all cases unless specifically stated otherwise, it will be the astragalus which is meant. (See Fig. 1.)

In animals in which the foot is developed to a certain extent, and in man himself, the astragalus is not symmetrical but has developed in order that the strain of the weight of the animal may be taken up by the foot. The fact that it is the symmetrical astra-

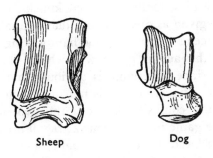

Sheep   Dog

Fig. 1—Animal astragali—natural size

galus which is found in quantity may be a fact of significance, although it is again conjectural as to what might have been the purpose of the collection. The bones may have been used as tally-stones for counting in the same way as the coloured stone pebbles. This is a distinct possibility, although it is more likely, considering their size, that they were used for games of some kind or perhaps as toys. They are pleasant things to handle and take a high polish with use. It is not uncommon to see children in France and Italy playing games with them today, and idealised versions of them made in metal can be bought in the village shops.

If primitive man prized his astragali for some magical reason, it is possibly not too fanciful to suppose that he would have carved or decorated them in some way, much as he did the larger bones such as the shoulder blade, but an inspection of a number of museums devoted to the relics of primitive man has not revealed any such decorated heelbone. All we may record is that finds of coloured pebbles and astragali are well known to the prehistorian, and we may conjecture from this fact that prehistoric man may have had the notion of counting and that he, or his children, or both may have used pebbles or astragali for toys.

Although the uses of the astragali in prehistoric times are a matter for conjecture, we find ourselves on firmer ground when we consider the Babylonians and the Egyptians, the Greeks and the Romans of the pre-Christian era. There is no doubt now that one of the uses of the astragalus is as a child's toy. We are told that students and schoolboys played knucklebones everywhere, that they were given as presents, and that one schoolboy received eighty all at once as a prize for good handwriting. It is difficult to deduce what the actual game was which the translators call knucklebones, since this name is obviously a transference from our own time. The astragalus is not unlike the knucklebone to look at, and the confusion between the two names may have arisen from the game in which four astragali are balanced on the knuckles of the hand, tossed and caught again. Since the origin of children's games is often lost in antiquity it may well be that the classicists are correct in calling the game played by the ancients by its modern name. From the evidence of Greek vase paintings, however, it would seem just as likely that the game was one where a ring was drawn on the ground and the players tossed the astragali into it in much the same way as the children of our own time play marbles. What is not in doubt is the universality of the game. According to Homer, when Patroclus was a small boy he became so angry with his opponent when playing a game of knucklebones that he fought and nearly killed him ; there are many other such references.

Whether the child aped the man or the man adopted the toys of his children it is not possible to say, but the astragalus was certainly in use for board games at the time of the First Dynasty in Egypt (c. 3500 B.C.). For those interested in the development of board games the evidence of this period provides many puzzles. There is the archaeological evidence, that is the actual boards and the counters which are dug up with them. There are also the tomb-paintings in which one stage of the game is depicted. Some of these games, such as those played by the workmen who built the Pyramids, bear some resemblance to draughts or to noughts and crosses. Others may have been a primitive form of halma, chicken or tic-tac-toc, but in some games undoubtedly the counters or " men " were moved according to some established rule and after the tossing of an astragalus. In one Egyptian tomb-painting a nobleman is shown playing a game in his after-life—that being his favourite recreation in this life. A board is set out with " men " in front of him and an astragalus is delicately poised on a finger tip (Plate 1, facing page 8). A game which the excavators call " Hounds and Jackals " is shown in Plate 2 (between pages 8 and 9). The hounds and jackals were moved according to some rule by throwing the astragali found with the game and shown in the illustration. There are many boards which bear a similarity to this one which is intended perhaps for a game such as " Snakes and Ladders", still played today and which is itself a primitive form of backgammon.

Even earlier than these stylised board games, we have from a report of excavations in Turkey the suggestion that the astragalus and pleasurable recreation are linked together. A wine-shop of the thirteenth century before Christ is described as follows*:—

> On the inner side of the room were the ruins of a brick structure resembling a bar : behind it were two partly sunk clay vats, and in the corner a pile of ten beautiful chalices of the " champagne " type. There was much other pottery in the room. Near the door was a pile of 77 knucklebones, and beside them, perhaps used as tokens in a game, a stack of the crescent-shaped objects, till now usually described as loom-weights. The remainder of the floor space was mainly occupied by skeletons.

It is possible, but not altogether likely, that these games originated in Egypt. On the whole, like the concept of number, the balance of probabilities would appear to be in favour of their having a still older origin and to have spread westwards from Arabia, or India, or even farther east. It is most unlikely that the origin was Greek, although Herodotus, the first Greek historian, writing about 450 B.C., is willing to credit his fellow countrymen or allied peoples with inventing much. Concerning the famine in Lydia (c. 1500 B.C.) he writes†:

---

* Quoted from *The Times*.
† *History* (tr. by George Rawlinson), Everyman's Library, Vol. I, page 96 (J. M. Dent & Sons Ltd. & E. P. Dutton & Co. Inc.).

The Lydians have very nearly the same customs as the Greeks. They were the first nation to introduce the use of gold and silver coins and the first to sell goods by retail. They claim also the invention of all games which are common to them with the Greeks. These they declare they invented about the time that they colonised Tyrrhenia, an event of which they give the following account. In the days of Atys, the son of Manes, there was great scarcity through the whole land of Lydia. For some time the Lydians bore the affliction patiently, but finding that it did not pass away, they set to work to devise remedies for the evil. Various expedients were discovered by various persons ; dice* and hucklebones and ball and all such games were invented, except tables (i.e. backgammon), the invention of which they do not claim as theirs. The plan adopted against famine was to engage in games on one day so entirely as not to feel any craving for food, and the next day to eat and abstain from games. In this way they passed eighteen years.

In yet another commentary we are told that Palamedes invented games of chance during the Trojan Wars. During the ten years investment of the city of Troy various games were invented to bolster up the soldiers' morale, since they suffered from boredom.

The die had undoubtedly been evolved by this time, as I shall later observe, but in classical times the principal randomising agent (to use the modern jargon) would seem to have been the astraga us. We shall never know the genius who first introduced the random element in this way. The conjunction of the coloured pebbles and astragali on the prehistoric sites and of the coloured counters and astragali in the early board games is suggestive (and tantalising), but there will probably never be enough evidence to link the two. One may conjecture that since both pebbles and astragali might have been concerned with counting and that since

---

* The Greeks called the die *tessera*. The word comes from the Greek for " four " and the reference is to the four edges of one side of a die. The distinction between the die and the astragalus is not one which is often made, translators and commentators usually translating astragalus as die, while the historians themselves sometimes write *talus* (the heelbone) instead of *astragalus*.

the astragalus has four flat sides and is pleasant to toss, the development of the board games came in this way, but it is only conjecture. That *gaming* was developed from *game-playing* is however at least a possibility. Although game-playing was highly developed in Egypt *c.* 3500 B.C. gaming is said to have been introduced with Ptolemy from Greece only 300 years before Christ. It may be suggested accordingly that gaming could be a further idealisation of the board game where the random element is retained and the board dispensed with. It is, however, equally likely that gaming developed from the wager and the wager from the drawing of lots, the interrogation of the oracles, and so on, which have their roots deep in religious ritual.

By the time of the emergence of Rome and the Romans as the dominating power in Europe, gaming was the common recreation among all classes and types of people, so much so that it was found necessary to promulgate laws forbidding it, except at the Saturnalia. It would not appear that these prohibitions were very effective since they were repeatedly renewed, and ignored. What game or games were played by the common man we do not know, although it would seem likely that whatever they were the astragalus played a prominent part. Much was written in the Middle Ages about this gaming with huckle (or knuckle) bones. The astragalus has only four sides on which it will rest, since the other two are rounded (see Plate 3, between pages 8 and 9). A favourite research of the scholars of the Italian Renaissance was to try to deduce the scoring used. It was generally agreed from a close study of the writings of classical times that the upper side of the bone, broad and slightly convex, counted 4; the opposite side, broad and slightly concave, 3; the lateral side, flat and narrow, scored 1, and the opposite narrow lateral side which is slightly hollow, 6. The numbers 2 and 5 were omitted.

Four astragali were often used in a simple " rolling of the bones". From a tossing of a modern sheep's astragalus the empirical probabilities were approximately one in ten each of throwing a 1 or of throwing a 6 and about four in ten each, of throwing a 3 or a 4, so that the probabilities associated with the four astragali would have been—

Throw*				$10^4 \times$ Probability for each throw.
(1^4)	(6^4)			1
(3^4)	(4^4)			256
(1^33)	(1^34)	(6^33)	(6^34)	16
(3^31)	(3^36)	(4^31)	(4^36)	256
(1^36)	(6^31)			4
(3^34)	(4^33)			1024
(1^23^2)	(1^24^2)	(6^23^2)	(6^24^2)	96
(3^24^2)				1536
(1^26^2)				6
(1^234)	(6^234)			192
(3^216)	(4^216)			192
(3^214)	(3^264)	(4^213)	(4^263)	768
(1^263)	(1^264)	(6^213)	(6^214)	48
(1634)				384

The " one " was called the dog by Greeks and Romans or sometimes, less frequently, the vulture. The best of all throws with four bones was called the " venus " when all the sides uppermost were different. But there is no reason to suppose that the method of scoring was always the same or that some special throws did not count more than others. The throw of Euripides with four astragali is said to have been worth 40. How the bones fell to achieve this result does not appear in the classical literature, but Cardano, writing in the sixteenth century, says it was four 4's. The fact that 40 is approximately the inverse of the probability of achieving this result is, one would think, coincidental.

In the *Lives of the Caesars* Suetonius (*c*. 69 – *c*. 141) makes several references to the passion for gaming which possessed the Emperors. The prohibition on such games except at the Saturnalia was clearly, by this time, disregarded. Thus in the " Life of Augustus " (63 B.C.–A.D. 14)—we read†:

---

* I use here the common partitional notation. Thus (1^4) means one uppermost on each of the four bones, (1^264) means one uppermost on each of two bones with a six and a four uppermost on the other two, and so on.

† Translated by J. C. Rolfe, Loeb Classical Library, by courtesy of Wm. Heinemann Ltd. and Harvard University Press.

Egyptian tomb painting showing a nobleman in after-life using an astragalus in a board game (*see page* 4)

(*By courtesy of the Oriental Institute, University of Chicago*)

Plate 1

The board game of "Hounds and Jackals" (*c.* 1800 B.C.), **from the
tomb of Reny-Soube at Thebes, Upper Egypt** (*see page* 4)

The wooden stand is overlaid with ivory and ebony

(*By courtesy of the Metropolitan Museum of Art, New York*)

Plate 2

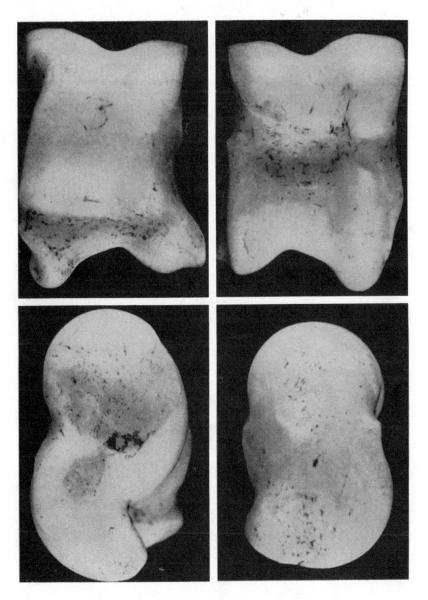

Four views of an astragalus (*see page* 7)

Plate 3

Die roughly made on classical pattern

Imitation astragalus carved in stone

Egyptian carved ceremonial die (*centre*) **with typical astragali**
(*see page* 10)

Plate 4

He [Augustus] did not in the least shrink from a reputation for gaming and played frankly and openly for recreation, even when he was well on in years, not only in the month of December, but on other holidays as well and on working days too. There is no question about this, for in a letter in his own handwriting he says : " I dined, dear Tiberius, with the same company ; . . . We gambled like old men during the meal both yesterday and today, for when the dice were thrown whoever turned up the dog or the six put a denarius in the pool for each one of the dice, and the whole was taken by anyone who threw the venus."

(It would be appropriate here to read *astragali* for *dice*.) Again in a letter to his daughter we read :

I send you 250 denarii, the sum which I gave each of my guests in case they wished to play at dice or at odd and even during dinner.

There are several other references to Augustus of this nature.

Concerning Claudius (10 B.C. – A.D. 54) Suetonius is not so explicit. He remarks that :

He was greatly devoted to dicing, even publishing a book on the art, and he actually used to play while driving, having the board fitted to his carriage in such a way as to prevent his game from being disturbed.

It is unfortunate that this book of Claudius on " How to win at Dice " has not survived. However, it is likely that it would not have contained directions on how to play the games of chance of his day, but rather that it was an exposition on the theme that the only way to win games of chance is to have a genuine wish to lose them, much in the vein of the old story where " a man is promised 1000 gold pieces every time he meets a stranger riding on a piebald mule, but only on condition he does not think of the mule's tail until he gets the money".*

The transition from the astragalus to the die probably took place over thousands of years, and it is possible that the first primitive dice were made by rubbing the round sides of the astragalus until they were approximately flat. There are many of these osselots in existence, and they cannot have been entirely satisfactory since the honeycomb tissue of the bone marrow has

---

* I have taken this simile from Robert Graves' *I, Claudius*.

been exposed in some. Most of them have a letter or letters cut deeply on one side. There are also imitation osselots cut out of solid bone. These are roughly cubical, but the carver has cut out the hollows in similitude of the natural astragalus.

The earliest dice so far found are described as being of well-fired buff pottery and date from the beginning of the third millennium. A die from Tepe Gawra (N. Iraq) has the opposite points in consecutive order, 2 opposite 3, 4 opposite 5 and 6 opposite 1 (Fig. 2 (i)). The Indian die (Mohenjo-Daro) has again the opposite points in consecutive order but this time 1 is opposite 2, 3 is opposite 4, and 5 is opposite 6 (Fig 2 (ii)). That pips were used instead of some other symbolism probably just reflects the fact that at that time there would be no convenient symbolism

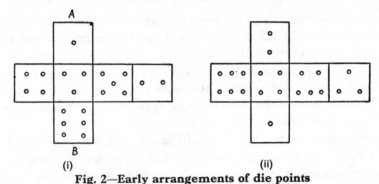

(i)   (ii)

Fig. 2—Early arrangements of die points

for number and that dots were easier to make than straight-line cuts. This consecutive order of the pips must have continued for some time. It is still to be seen in dice of the late XVIIIth Dynasty (Egypt c. 1370 B.C.), but about that time, or soon after, the arrangement must have settled into the 2-partitions of 7 familiar to us at the present time. Out of some fifty dice of the classical period which I have seen, forty had the " modern " arrangement of the pips.

The classical dice vary considerably in the materials of which they are made and in the care with which they have been fashioned. The impression left by many of them is that the maker picked up any convenient piece of stone, or wood, or bone and roughly shaped, marked and used it (see Plate 4, facing page 9).

This is possibly not surprising since even these imperfect dice must have seemed good after the astragali. There are exceptions. A few of the dice I have seen are beautifully made, with tooled edges, and throw absolutely true. In classical times we have also the imitation astragalus cut from stone and decorated with carvings (Plate 4). There are also many variants of lewd figures fashioned in metal or bone. These figures could fall in six ways, have dots from 1 to 6 cut in them—although all positions are not equally probable—and were obviously used for gaming.

The primitive dice which have been found in excavations in England might also be mentioned. These date from the invasion of England by the Romans and are possibly part of the benefits spread by that civilisation. The longbone of an animal such as a deer was taken and shaved so that it was reasonably square in

Fig. 3
Hollow die
shaped from
bone

cross-section (Fig. 3). A piece of bone was then cut so that it formed a cube. The resulting die must obviously have been very unsatisfactory since it was a hollow square cylinder with 1 opposite 6, 2 opposite 5 on the solid sides and 3 opposite 4 on the hollow ends, the marks being made with a circular engraving tool. Sometimes the die has a 3 on each of the hollow ends.

The idealisation of the astragalus by the die possibly took place simultaneously in different parts of the world, and the results were not always cubical. Throwing-sticks of the order of 3 inches long and roughly 1 cm. in cross-section were used by many peoples, among them the ancient Britons, the Greeks, the Romans, the Egyptians and the Maya Indians of the American continent. Sometimes the cross-section is a square, sometimes it is elliptical. They are all alike in having only four numbers on them, one on each side of a longitudinal face and the other two at either end of the opposite longitudinal face. The numbers, in contrast to the astragalus, are mostly 1, 2 : 5, 6, though 3 and 4 have been noticed. These are most often marked by cuts, though some are with pips.

The Greek mathematicians worked out the geometry of regular solids in the latter half of the first millennium, and this was probably followed by the construction of polyhedral dice. The beautiful icosahedron made of rock crystal, in the Musée du Louvre (Fig. 4), is one of these. The sides are numbered in Roman figures one to twenty, and there are also letters engraved, following perhaps the Greek custom of associating letter with number.* The theorems concerning the regular solid had not, however, apparently spread very far because there was produced about this time a solid with 19 faces, supposedly rectangular, a die with 18 faces possibly formed by beating out a cubical die, one

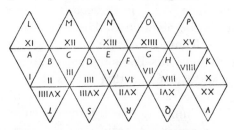

Fig. 4—Arrangement of icosahedral die

with 14 faces and so on. There are also a certain number of faked dice from this period. Apart from the device of leaving out one number and duplicating another, an early die has been described in which the " one " side can be lifted to expose a hollow underneath. It is suggested that a small ball of leather could be crammed into this to alter the balance of the throw.

The beginning of the Christian era finds us then with dice, with astragali, with throwing-sticks, with board games, and with games of chance which use neither boards nor men. The idea of counting and enumeration is firmly established but not the concept of number as we know it now. The paraphernalia of chance events has been organised for man's pleasure and entertainment. Randomisation, the blind goddess, fate, fortune, call it what you will, is an accepted part of life. But for an understanding of man's mental attitude towards these chance events, and his conception of chance in general, it is necessary to turn attention to a different stream of thought—divination.

---

* The Greek system of numbering c. 200 B.C. for what we would now write as 1, 2, 3, 4, 5, 6, 7, 8, . . . was α, β, γ, δ, ε, σ, ζ, η.

chapter 2

Divination

This is the third time ; I hope good luck lies in odd
numbers. . . . There is a divinity in odd numbers, either
in nativity, chance or death.

Sir John Falstaff, *Merry Wives of Windsor*, v. i. 2.

A fundamental in nearly all religions is some sort of mechanism
whereby the deity may be consulted and if willing make his (or
her) wishes known to the suppliant. Some anthropologists see in
sortilege, or divination by mechanical means, the origin of many
simple games of chance. This is a " chicken or the egg " kind of
problem and it is not really possible to decide whether recreational
chance or divination by chance came first. It is, however, a
definite possibility that the equating of the intentions of the god
with respect to a suppliant with the random element of the game of
chance, surrounded that random element with a certain magic
which might have had the effect of rendering speculation about it
impious for many persons.

The simplest application of the random element for divination
purposes may be that of drawing lots or its equivalent. There are
many references to this among the social histories of primitive
tribes, where the detection of the guilty is accomplished by the
drawing of marked pieces of wood or straws of unequal length.
In Ezekiel xxi the correct path for an army to tread is determined
by four bowmen standing back to back and loosing off their arrows.
The early Teutonic tribes employed the drawing of lots in some of
their religious rites. The word " lot " is itself of Germanic origin
and found its way into the Romance languages through the
Italian " lottery " and returned to English by this route. The

13

casting of lots persisted for many hundreds of years. Close to our present time and in the memory of the writer the pious countryman was apt to treat the Bible as a book of random numbers, and a little more than two hundred years ago John Wesley sought guidance in almost this way. The reference is to his Journal of March 4th, 1737, and the point to be decided was whether he should marry or not.

> Having both of us [Mr. Delamotte and himself] sought God by deep consideration, fasting and prayer, in the afternoon we conferred together but could not come to any decision. We both apprehended Mr. Ingham's objection to be the strongest, the doubt whether she was what she appeared. But this doubt was too hard for us to solve. At length we agreed to appeal to the Searcher of Hearts. I accordingly made three lots. In one was writ, " Marry " : in the second " Think not of it this year". After we had prayed to God to " give a perfect lot", Mr. Delamotte drew the third, in which were the words, " Think of it no more". Instead of the agony I had reason to expect I was enabled to say cheerfully " Thy will be done". We cast lots again to know whether I ought to converse with her any more, and the direction I received from God was " Only in the presence of Mr. Delamotte".

In contrast to the " perfect lot " of the Journal, a " bad lot " ensued when enough prayer had not been made to enable God to show his will. This was presumably the way out which is found necessary in all these procedures to allow for cases where the result of the sortilege is not pleasing.

Wesley's sect was unusual among Christians in retaining sortilege. The Catholic Apostolic Church had condemned sortilege as being a relic of paganism, although it had been common practice at one time. The *Chronicle of Cambrai* records a dispute between the Bishops of Arras, Autun and Poitiers for the body of St. Léger which was resolved by sortilege when the lot (and the body) fell to the Bishop of Poitiers. Nowadays one might regard this as rather a sporting method of resolving a religious conflict, but this was early in the history of the Church and was undoubtedly a relic from the classical ages when important offices and vacancies in the hierarchy of the priesthood were

filled by lot. A variant of the sortilege was the " magic pictures."
A skin was stretched tightly over a ring and painted. Rings were
placed on the skin which was then rubbed. The dancing of the
rings over the pictures was then interpreted.

The game of odds and evens is one which has been used for
divination purposes from the first recordings of history and is still
used today among the African tribes. There are various cere-
monials attached to it. Some peoples use pebbles, some nuts, some
grain. Some pour from the horn of the buffalo, others from the
hands of their priests. If the random number of objects poured
out is odd some say the wish is granted, whereas others say it is
granted only if the number is even. But whatever the rules the
procedure is the same ; the question is posed, the lot is cast, the
answer is deduced. The method clearly does not give the god
much scope for self-expression but at least it produces an un-
equivocal answer.

The four astragali of the gambler were customarily used in the
temples of classical Greece and Rome, and the ceremony was
much the same as for odds and evens. The forecaster entered the
temple, stated his wish, picked up the four astragali (or the
attendant votary did for him) and cast them on the table. From
the four uppermost sides the answer was deduced, usually from
reference to the tablets of the oracle on which the throws and their
meanings were set out. In this probing of the future the venus
(1346) was a favourable omen (probability $c.$ 0·04) while the dogs
(1^4) (probability $c.$ 0·0001) was unfavourable. The fact that 2
and 5 do not seem to enter into prediction for the Greeks and the
Romans would probably indicate that dice were not used.

While there seems some unanimity that only four astragali
were used in Greece and in Rome, inscriptions in Asia Minor
indicate the use of five, or possibly one astragalus used five times.
(Remembering the difficulties of arithmetic and of the noting
down of the number thrown, it is likely that five bones were used
at one throw rather than one bone five times.) Sir James Frazer
in his *Commentary on Pausanias* has translated some of the inscriptions. Each throw was given the name of a god, but whether the
order in which the astragali fell mattered we shall probably never

know for certain. The translation of the throws of the saviour Zeus (probability *c.* 0·077), of the Good Cronos (probability *c.* 0·013), of Poseidon (probability *c.* 0·013), and of child-eating Cronos (probability *c.* 0·006), are given by Sir James Frazer as follows :

>1.3.3.4.4. = 15. *The throw of Saviour Zeus.*
>>One one, two threes, two fours,
>>The deed which thou meditatest, go do it boldly.
>>Put thy hand to it. The gods have given these favourable omens.
>>Shrink not from it in thy mind, for no evil shall befall thee.
>
>6.3.3.3.3. = 18. *The throw of Good Cronos.*
>>A six and four threes.
>>Haste not, for a divinity opposes. Bide thy time.
>>Not like a bitch that has brought forth a litter of blind puppies.
>>Lay thy plans quietly, and they shall be brought to a fair
>>>completion.
>
>6.4.4.4.4. = 22. *The throw of Poseidon.*
>>One six and all the rest are fours.
>>To throw a seed into the sea and to write letters,
>>Both these things are empty toil and a mean act.
>>Mortal as thou art, do no violence to a god, who will injure thee.
>
>4.4.4.6.6. = 24. *The throw of child-eating Cronos.*
>>Three fours and two sixes. God speaks as follows :
>>Abide in thy house, nor go elsewhere,
>>Lest a ravening and destroying beast come nigh thee.
>>For I see not that this business is safe. But bide thy time.

One might hazard a guess and say that as the numbers are mentioned not necessarily in ascending order, this would indicate that there was to be a definite order in the fall of the astragali and in the interpretation of the throw. Reflection however prompts the thought that 4^5 tablets would be necessary if all eventualities were to be covered. This oracle seems to have been kinder than many in that the throw of child-eating Cronos is more difficult to attain than the others.

According to Waddell,* something of the same system was practised in Tibet at the end of the last century, this time re-

---

* L. A. Waddell, *The Buddhism of Tibet* (1894).

garding the prediction of the next stage of reincarnation. The prediction is reported as being carried out by a priest. One rebirth chart consists of 56 two-inch squares arranged 8 × 7. Each square corresponds to a future state. A six-sided die with letters on it instead of pips is thrown down on the rebirth chart, and according to the square on which it lands and the letter which falls uppermost so the prediction is made. Waddell, who visited Tibet (c. 1893) as a member of a British Mission, obtained one of these charts and a die. He remarks "the dice [sic] accompanying my board seems to have been loaded so as to show up the letter Y, which gives a ghostly existence, and thus necessitates the performance of many expensive rites to counteract so undesirable a fate". It has been noted that some of the ancient osselots which had been roughly squared off by rubbing had letters deeply engraved on them, and the letter Y is included among these. These may, accordingly, have been used for ceremonies similar to those with the Tibetan dice. On the other hand the fact that letters instead of pips are used on either die or osselot may merely reflect the interchangeability of letter and number symbol which is so marked a feature of the early struggles in writing down numbers.

In addition to the divination carried out in the temples in classical times, with or without the aid of the attendant votary, it was a commonplace for individuals to perform these searchings as part of their daily lives. Suetonius tells us that Tiberius, before he became emperor, would seek for omens by throwing golden tali into the fountain of Aponus close to the hills of Euganea. He remarks, rather improbably, that he had been there and that the tali could still be seen in the water. But it was not only Tiberius : Propertius bursts out with " when I was seeking Venus with favourable tali the damned dogs always leapt out". The witty Lucian, telling the story of the young man who fell madly in love with Praxiteles' Venus of Cnidos, writes :

> He threw four tali on the table and committed his hopes to the throw. If he threw well, particularly if he obtained the image of the goddess herself, no two showing the same number, he adored the goddess and was in high hopes of gratifying his passion : if he threw badly, as usually happens, and got an unlucky combination,

he called down curses on all Cnidos, and was as much overcome by grief as if he had suffered some personal loss.

This type of personal soothsaying also seems to have survived to the present time, since it is reported that some laymen in Tibet are equipped with a pocket divination manual for interpretation of the omen produced by some sort of lot-casting. I have been told also that some business men in this twentieth-century England consult an astrologer before making a major decision.

It is possible to speculate that sortilege spread westwards from India and Tibet to the Mediterranean, although if this is so it is curious that divination there by dice was apparent only after the Greek infiltration. The beautiful cubical die shown at the lower centre of Plate 4 (facing page 9), cut from hard brown limestone with sides of about one inch, and having the sacred symbols of Osiris, Horus, Isis, Nebat, Hathor and Horhudet engraved on it, was possibly used for divination purposes and dates from the Ptolemaic dynasty (300–30 B.C.) ; I have not been able to trace an earlier specimen. It has been stated that games of chance with dice did not penetrate to Egypt until the arrival of Ptolemy, and probably soothsaying with dice came at the same time. On the other hand, moving the " men " of board games with astragali, as we have noted, was common in Egypt, and it is likely, although there is no record of it, that astragali were also used for divination purposes.

Two astragali were, apparently, often used in the Egyptian board games. Four and sometimes five astragali were used in the Roman temples, but when dice began to be used for prediction three seems to have been the preferred number, although two were still used for games of chance such as hazard or backgammon.* This may have been because the rules for these games were well established universally and would have been difficult to alter. With divination, however, it may not be too fanciful to see the mystic number 3 of the Christian era—the beginning, the

---

* Robert Graves in *The White Goddess* mentions this point. I think, following his theory, he would say that only eighteen sides were necessary for alphabetic divination. I am not sure that this, however, is not unduly restricting the divination.

middle and the end—body, soul, spirit—spirit, water, blood—the trinity—ousting the sacred number 4—earth, air, water, fire— and 5 which was the basis of all number. Pythagoras taught that three was the perfect number, the symbol of all deity, and the new Christian sect appropriated his idea. The augurers possibly followed suit.

Divination by the individual lasted well into the Christian era. In fact, as the pagan world disintegrated, rule by divination increased and sometimes attained monstrous proportions. Septimius Severus (Emperor A.D. 193–211) was notoriously at the mercy of every omen and portent. Diocletian (Emperor A.D. 284–305) was no better. Eventually Christianity swept away much of public and private oracularity and turned the rest into satanic inspiration, but it did not abolish divinatory practices altogether. During the drums and tramplings of four thousand years, the astragalus and the die survived, the latter slowly (but very slowly) supplanting the former. The birth of Christ thus marks an interesting and puzzling situation. In spite of the sophistication of manners and morals achieved at this time, man was still a child in his conceptual thought, and instead of (or perhaps as well as) the conceptual idealisation of the Unknown God, familiar gods and goddesses lurked under every stone of his path. To appease one was to offend the other, and the constant recourse to lot-casting, tali, and so on to probe the divine intent was a solution of a difficulty for which one has every sympathy. What is puzzling is that the same ritual was used in the taverns, eating-houses and so on, to while away the idle hour, and there does not seem to be an answer to this.

Under the withering attack of Christianity the pagan gods dwindled into fairies, their fertility rites into dances round the maypole, their sacrifices into furtive libations to the little men. But such things do not die suddenly or spectacularly. They weaken and fade but they endure with the human desires from which they sprang.* All through the ages we find chance

---

* It is still possible to get a horoscope cast in London. It is illegal to foretell the future, but it is not illegal to express an opinion about the attitude of the stars towards the prospect of an enterprise.

mechanisms—dice in particular in our millennium—used to interrogate the unknown. One of the best-known medieval poems —its survival alone is some indication of its popularity—is " The Chance of the Dyse " which lists all the 56 possible throws with three dice. This is definitely a fortune-telling poem with an interpretation given for each throw. Thus, for example :

> Thou that hast six, five, three
> Thy desire to thy purpose may brought be.
> If desire be to thee y-thyght
> Keep thee from villainy day and night.*

It is a little difficult to imagine our hardboiled countrymen of the Wars of the Roses really determining their course of action from the fall of the dice, but possibly this was an ancillary performance which some went through to be on the safe side.

Playing-cards, as far as one can judge at present, were not invented until *c.* A.D. 1350, but once in use, they slowly began to displace dice both as instruments of play and for fortune-telling. Early cards were expensive and beyond the means of the average gambler, and in consequence it was some hundreds of years before they finally ousted dice. Their influence on the early history of probability was, accordingly, slight. Huygens and James Bernoulli obviously thought of the random element in terms of dice. The mixture of dice problems and card problems really began with Montmort and de Moivre. Dice-playing has now almost disappeared from the Western world except in games for children and for the crap-shooting citizens of the U.S.A., but cartomancy in many forms has survived.

---

* It would be of interest to try to match these prognostications against those of the " caves " of the oracles. Superstitions in religious guise are often self-perpetuating.

chapter 3

The probable

Egypt's might is tumbled down,
Deep a-down the deeps of thought ;
Greece has fallen and Troy town,
Glorious Rome hath lost her crown,
Venice' pride is nought.

But the dreams her children dreamed,
Fleeting, unsubstantial, vain,
Shadowy as the shadows seemed,
Airy nothing, as they deemed,
These remain.

MARY COLERIDGE

We may speculate as we please about number, about the rules of the various games of chance, about the use and misuse of the religious auguries, but there is no denying that the real problem which confronts the historian of the calculus of probabilities is its extremely tardy conceptual growth—in fact one might almost say, its late birth as an offspring of the mathematical sciences. For it was fifteen hundred years after the idealisation of the solid figure before we have the first stirrings, and four hundred and fifty years after that before there was the final breakaway from the games of chance, if indeed we have really accomplished it even today. The random element was introduced before recorded time, and there are enough references in the literature before the birth of Christ to indicate that this random element—surely the goddess Fortuna herself (Plate 5, facing page 24)—was pursued with assiduous fanaticism. Why then was the concept of the equally-likely possibilities in die-throwing so long delayed? It is a question which one must, perhaps, try to answer in terms of the emotions rather than the intellect.

So long as bones were used for play and divination, the

regularity of the fall of the different sides of the osselot would be obscured. It would undoubtedly be affected by the kind of animal bone used, by the amount of sinew left to harden with the bone, by the wear of the bone. All these, empirically, can be shown to alter the probability of any one side coming uppermost to a greater or less extent. One must add to this the fact that a long series of trials would be necessary to calculate the empirical probabilities and that there would be few persons who, even if they wanted, would be capable of keeping the tally of the throws and of making the necessary enumeration. For, as has been previously noted, in classical times the writing down of numbers was a difficulty. The transition from osselot to die was accompanied by large numbers of imperfect dice. This is understandable, for if there was no regularity with the astragalus, and since the idea of the perfect solid was newly created, then any handy piece of wood or ivory or stone, smoothed off and marked, would seem appropriate. It is likely therefore that the common citizen at the beginning of the Christian era had no realisation of what was later called " the stability of statistical ratios".

But it must also be remembered that the dice of 3000 B.C. are of hard, well-fired buff pottery and would have played true, while some of the classical dice are beautifully made and show no bias whatsoever. Some of these dice, at least, were used by the priesthood in divination rites and the priesthood would, on the whole, have tended to be the educated class. Further the die would have been cast so often in the religious ceremonies that considerable experience would be available ; it is curious that the conceptual breakthrough does not come at this point shortly before the birth of Christ. Several explanations occur to one, none of them convincing, as to why the regularity of the fall of the dice was not noticed.

It is just possible that during his novitiate the priest was taught to manipulate the fall of the dice, or even of the astragali, to achieve a desired result. This being so there would be no need to calculate probabilities or even to consider the relative values of the different throws because the random element would be missing. This would be in keeping with the attitude of the priest as the inter-

preter of the divine intent. Another possibility is that the calculation of these probabilities was in fact carried out but that it was
part of the mystery of the craft and not divulged. This is not
so likely as the first, for with the changeover to Christianity
someone would undoubtedly have talked. A third and more
plausible possibility is that speculation on such a subject might
bring with it a charge of impiety in that it could be represented
as an attempt to penetrate the mysteries of the deity. At first sight
this explanation is also a little improbable. Priests and philosophers are not usually deterred by such considerations. Priests,
honest and sincere men, have a habit of treating the mysteries of
their craft with an easy familiarity which the agnostic finds
startling, while everything is grist to the mill of the philosopher.
Nevertheless this reluctance to probe may have been the biggest
stumbling-block to the bridling of the random element in its
simplest form.

It is to the Greek philosophers that we owe the development of
scientific logic, which was to exert its baleful influence on the
natural sciences for nearly fifteen hundred years, and it is to these
same logicians that we owe the idea of the enumeration of
possibilities or causes and some suggestion that the probability
of an event is not entirely subjective. The hunting dog, used by
Chrysippus to illustrate the enumeration of possibilities, is well
known to logicians. A dog chasing his quarry arrives at a place
where the road splits into three. He tries to find the scent along
two of the paths and then continues along the third without
making a cast. Thus if there are a fixed number of possibilities
and all but one are discarded the one left must be the right one.

It was Chrysippus also who tried to calculate the correct
diagnosis of an illness by discussing the number of " molecular
propositions " which arise from all possible combinations of the
" atomic propositions". The idea of combinations seems to have
been a familiar one to the Greek philosophers* who were hampered

* S. Sambursky (reference q.v.) quotes Plutarch as writing that " Xenokrates
determined the number of syllables which are produced through mixing
the letters of the alphabet up to 1,002,000 millions".

possibly by the difficulty of writing down numbers.　Until this difficulty was resolved, and it never was by the Greeks, the necessary algebraic development could not come.

Not only do we have the first fumblings towards the enumeration of possible events and the first stirrings of combinatory theory but we also find, in the Greek mathematicians and the Roman philosophers, an alternative to the idea that the gods controlled the fall of the dice or the astragali in games of chance, although they were still wary of the religious omen. The suggestion is there of an instinctive feeling about probability which was never formulated.　Thus Cicero (*De Divinatione*), putting up an argument in order to destroy it, has Quintus say in Book I :

> They are entirely fortuitous you say? Come! Come! Do you really mean that? . . . When the four dice produce the venus-throw you may talk of accident : but suppose you made a hundred casts and the venus-throw appeared a hundred times ; could you call that accidental?

The inference is here that a run of one hundred venus-throws is so improbable that the god (or goddess) must be intervening to cause it.　Cicero in his reply in Book II demonstrates in his customary trenchant style that he has a firm grasp of the idea of randomness.

> Is it possible, then, for any man to apprehend in advance occurrences for which no cause or reason can be assigned? What do we mean when we employ such terms as luck, fortune, accident, turn of the die, except that we are seeking to describe events which happened and came to pass in such a way that they either might not have happened and come to pass at all or might have happened and come to pass under quite different circumstances? How then can an event be anticipated and predicted which occurs fortuitously and as a result of blind chance and of the spinning of Fortune's wheel?*

---

* Fortuna is often portrayed standing on a ball or a wheel, often blindfold. The wheel is usually introduced in pictures (cf. Plate 5, facing page 24) as a symbol of uncertainty or insecurity.

Lorenzo Leombruno's Allegory of Fortune (16th century) *(see page 21)*
The goddess is characterised as " Dea varia, lubrica et fragilis "

(By courtesy of Mansell—Alinari)

Plate 5

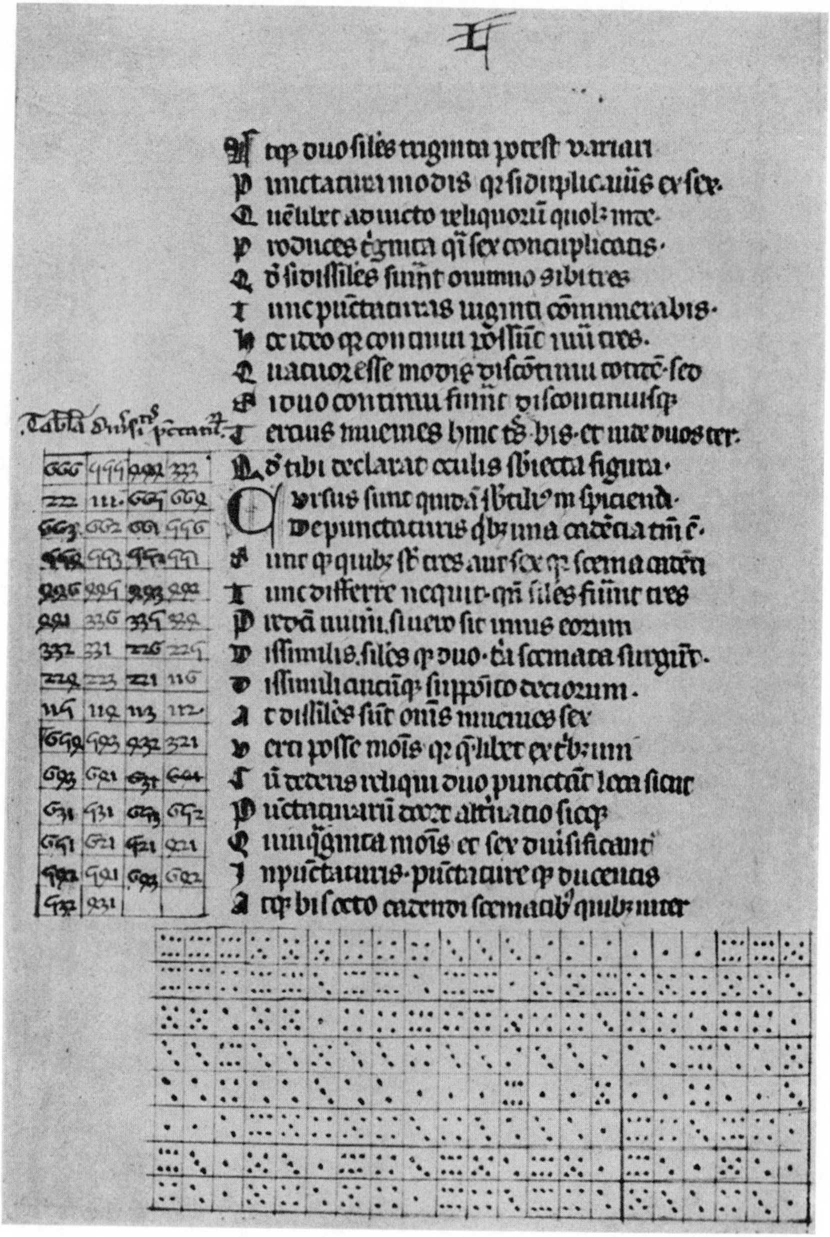

Plate 6

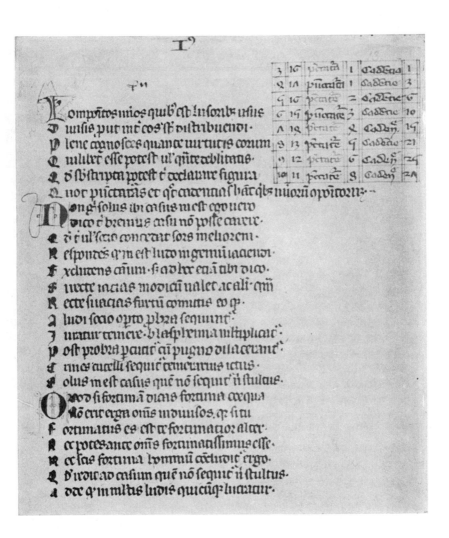

Excerpt from the manuscript poem "De Vetula" (Harleian MS. 5263)
giving the number of ways in which three dice can fall (*see page* 33)

(*By courtesy of Dr. M. G. Kendall and the Editor of "Biometrika"*)

Plate 7

Fra Luca Paccioli (National Galleries of Capodimonte, Naples)
(*see page* 36)

(*By courtesy of Mansell—Alinari*)

Plate 8

And again in castigation of soothsayers :

> Do you really feel that lots require any discussion? What is a lot anyway? It belongs virtually in the same category as "guess-the-fingers",* knucklebones and dice. In all these games audacity and luck win, not reason and thought. As a matter of fact the whole system of peering into the future by means of lots was the invention of tricksters who were only interested in their own financial welfare or in fostering superstition and folly.

This idea of chance persists through the whole essay. He writes again :

> Nothing is so unpredictable as a throw of the dice, and yet every man who plays often will at some time or other make a venus-cast : now and then indeed he will make it twice and even thrice in succession. Are we going to be so feeble-minded then as to aver that such a thing happened by the personal intervention of Venus rather than by pure luck?

We have then the idea that a run of venus-throws is improbable in that some seek an extraneous cause for its happening, the idea that if play is sufficiently prolonged the rare event will occur, and the concept of the random event. Further in his essay Cicero firmly divorces all prognostications and sooth-saying from interpretation of the will of the gods. His ideas would not have spread to the uneducated and illiterate and, possibly, would have been unpopular with the priesthood, but the awareness is there and the ingredients for the development of the concept of randomness seem complete.

It has been suggested by several writers that the reason why the next step was not made at this time was because the philosophers themselves, notably Plato and Aristotle, expected to find regularity, order and repetition of movement in the heavenly bodies but had no such expectation and therefore did not look for it in the events immediately surrounding them. It is more

---

* Possibly the same game still played by children today. One holds a hand behind his back and the other is required to guess the number of fingers folded.

likely perhaps that the step was not taken because of the gap between theory and practice which was to stultify scientific thought until the fermenting, fertilising flood of the Italian Renaissance. In other words I suggest that the step did not come at that time because the philosophic development which opened so many doors for the human intellect engendered a habit of mind which made impossible the construction of theoretical hypotheses from empirical data.

With the advent of Christianity the concept of the random event of the pagan philosophers was finally rejected. According to St. Augustine, for example, nothing happened by chance, everything being minutely controlled by the will of God. If events appear to occur at random, that is because of the ignorance of man and not in the nature of the events. Man's true endeavour was to discover and submit himself to the Divine Will, and not, presumably, to cloud this search by looking at patterns of behaviour in aggregates of events.

References

I would draw attention here to

S. SAMBURSKY, " On the Possible and Probable in Ancient Greece" published in the journal *Osiris* (1956), Volume 12, pp. 35–48. Sambursky gives a nearly exhaustive list of references relevant to early ideas about the " probable". Any edited library edition (with notes) of Cicero *De Divinatione* and the various classical dictionaries will also prove informative. Much remains to be done in the exploration of the role played by the random element in religious mysteries. I have used Hastings' *Encyclopaedia of Religion and Ethics* as a starting-point, but the field is vast. Robert Graves' *The White Goddess* is interesting about alphabetic divination, although not entirely relevant to my purpose.

chapter **4**

Early beginnings

> The wind bloweth where it listeth, and thou hearest the sound
> thereof, but canst not tell whence it cometh, or whither it goeth.
>
> ST. JOHN, iii. 8.

With the advent of Christianity and the decadence of the Roman
civilisation we enter the Dark Ages, which some date from the
collapse of Rome in A.D. 455. The Church became the haven of
all who cared for learning, and during the first years of this
millennium the early fathers thought that theology was the only
branch of scholarship which should be followed. The results and
achievements of Greek thought made themselves felt eastward, and
these years saw the flowering of Hindu mathematics and later the
rise of such centres as Alexandria. With the scholars who over
the centuries went east must have gone many of such manuscripts
as still existed of the classical world.

From the Hindus during the Dark Ages there came most of
what might be styled modern arithmetic and in particular the
writing down of numbers and the symbolism which we still use.
The symbol 5 for five, for example, is dated from *c.* A.D. 900*
when the Hindus introduced it, although it had to wait between
three and four hundred years before it appeared in Germany,
having passed through some vicissitudes on the way by the Arabs.
The notion that the position of digits in a number mattered we
get from India, although these mathematicians may have been
following the Babylonian tradition which hints at this. There are
two names which stand out from this time—Brahmagupta of the
seventh century and Bhaskara of the twelfth. Both these scholars

---

* Some would put it earlier.

27

wrote in Sanskrit verse and were concerned to summarise what
mathematical thought was known to the Hindus ; much of their
work is concerned with arithmetic and, in this arithmetic,
exercises on proportions. The problem of points which was to
become the focus of the attention of the sixteenth and seventeenth
century probabilists—" in what proportion should the stakes be
divided?"—has its origin in these exercises on proportions
though this early interest is probably not significant.* Such
examples as are quoted of their exercises indicate them to have been
concerned, principally, with the idea of fractions, possibly showing
here the influence of Nicomachus and his successor nearly four
hundred years later, Diophantes of Alexandria (born *c.* A.D. 325).

The Dark Ages for Europe were the golden ages for the Arabs.
Throughout the first millennium the Arabs were the dominating
peoples of the Mediterranean. They learnt from the Hindus and
they acquired the wisdom of the classical era by invading with
their armies and looting, among many other places, the libraries.
It was in this way (*c.* A.D. 641) that they obtained the fabulous
libraries of Alexandria with their collections of classical manuscripts. But although the impetus to conquer men exhausted
itself, the desire to conquer the problems of the intellect remained,
and it is to the Arabs that we owe the mingling of the ideas of
Hindu mathematical thought with the austerities of the Greek
philosophical method. From the Hindus they got number,
business arithmetic and astronomy, and from the Greeks they
got scientific logic. And being a seafaring race they developed
trigonometry and constructed astronomical tables. The important
Arab mathematician from many points of view is Mohammed ibn
Musa Al-Khowarizmi or Mohammed ben Musa or Alkarismi of
Khorassan who lived in the Caliphate of Al Mamum (*c.* A.D. 800).
He wrote two treatises, one on arithmetic and one on algebra,
both of which served as models and as treasure trove for the
mathematicians who were to come after him. It was probably his
use of the Hindu system of numbers which introduced it to

---

* Only a thorough examination would reveal if there was any examp
related to game-playing.

Western Europe. " The wondrous system of notation having 9 digits and a cipher, with device of place."

The first millennium in the countries of Western Europe was, on the whole, one of stagnation as far as the development of mathematical thought is concerned. It is the age of the wandering scholars who drifted from monastery to monastery, from centre of learning to centre of learning, speaking and writing their dog-Latin and disputing, mostly on theological problems, as they went. There were no frontiers for them and distance was no barrier to their wanderings, which provided the necessary fuel to keep the lamp of learning just burning. These migrations ended only with the formal founding of universities and the introduction of licences to teach at the beginning of the second millennium. But in the Dark Ages all was not entirely dark, although what light there was serves, in a way, to emphasise the darkness. It has already been mentioned that the Venerable Bede (A.D. 673–735) invented a system of counting, and he also wrote an arithmetic. The glory of the Dark Ages, for the present writer, undoubtedly belongs to his immediate follower and fellow north-countryman, Alcuin (A.D. 735–804) sometimes called Alcuin of York.* Apart from his beautiful haunting Latin verse, this great Englishman, both while at York and in his capacity as tutor to the sons of Charlemagne (A.D. 742–814), wrote textbooks on many subjects, including a book of arithmetical problems, *Propositiones ad acuendos Juvenes* (Problems for sharpening the mind of the young). He offered instruction in grammar, logic, rhetoric, arithmetic, music,

---

* Helen Waddell (*Mediaeval Latin Lyrics*) writes : " Alcuin, a Yorkshireman, died in his abbey of St. Martin at Tours in his seventieth year. In the spring of 801, three years before his death, he had written to his old friend the Archbishop of York with a little present of wine . . . and an entreaty that the Archbishop will not let his reading rust, *lest all my labour in collecting books be lost.* The Cathedral Library at York (the same which Bede must have used) had been Alcuin's passion ; he was Librarian and Master of the Schools there until Charlemagne persuaded him to Aachen." He wrote in his *Epitaphium:*

Alchuine nomen erat sophiam mihi semper amanti,
pro quo funde preces mente, legens titulum.

Alcuin was my name : learning I loved
O thou that readest this, pray for my soul. (Waddell's translation.)

geometry and astronomy.   Charlemagne is renowned for his foundation of universities, and it was Alcuin to whom was entrusted their organisation.   The Church, of which he was a devoted member, remembers him for his consistent and persistent advocacy of the idea that learning other than theology was not against its doctrines.

Throughout the Dark Ages the Church had, in spite of several liberal Popes, set its face against secular knowledge.   It had also even more vigorously fulminated against gaming.*   That it would disapprove of the sortilege in any form (dicing or lot-casting) is understandable since this was reckoned to be part of a pagan religion which, as the formalism of the new religion advanced, was something to be cast out.   But although the Church could, and did, dictate its own ceremonial, it was not successful in stopping gambling games.   The divines of the first millennium wrote and preached against games of chance, but the fact that they had to keep doing so is indicative of their failure to exorcise them. Possibly they were against gaming because of the vices which in their opinion accompanied it; perhaps it was just part of the general tendency of the Christian Church to achieve strength through self-denial; or again it may just have been the feeling that the gaming procedures were too closely interwoven with the pursuit of the old gods.   Whatever the reasons of the divines, they were against gaming, and at such a time when secular learning itself was sinful the idea that one might speculate on the chances involved would probably not be entertained.   Two noteworthy suggestions seem to stand out for the first millennium. First, if there was any formalisation of the concept of randomisation it would have been made in the Arab countries or in the lands further to the east.   Second, from the references in the literature and from the contra-activities of the Church, gaming flourished tremendously, and since the dice by this time would be well made the inveterate gambler, at least, must have had some intuitive idea of the empirical probabilities involved.   It will be noted later

---

* c.f. the Pardoner's remark:—
  This fruit cometh of the bitchéd bones two :
  Forswearing, ire, falseness, homicide.

that *c.* A.D. 1600 gamblers could detect a difference in probabilities of about 1/100 without being able to demonstrate the reason for it.

At the end of the first millennium and in the beginning years of the second the power of the Arabs declined and the reconquest by the Christian powers of lands held in fee by the Arabs began. Also the foundation of many universities took place between A.D. 1000 and 1200. There was thus almost simultaneously the spread of Arab learning and the means for improving on it. The sacking of Constantinople in 1453 released a further flood of manuscripts which found their way westward, and the stage was set for the Italian Renaissance. The forerunner of the revival of mathematical thought was possibly Leonardo the Pisan (A.D. 1170–1250), son of Bonacci, and sometimes called Fibonacci. The Italians had never ceased to be adventurers, and during the years after A.D. 1000, when the cities banded together for purposes of defence, commerce both inland and overseas flourished largely. We catch an echo of this in the well-developed insurance practice for merchant vessels by the Florentines, but it seems at that time a commonplace for both old and young merchants to seek trade eastwards. Bonacci was an Italian who was posted in Barbary— it is suggested as a kind of consular official—to look after the trading interests of the merchants of Pisa. His son Leonardo travelled extensively in the east, where he came into contact with the mathematical ideas of the Hindus and Arabs. Returning home he wrote the *Liber Abaci*, the first real mathematical treatise. In addition to giving the solution of many mathematical problems he expounded the Hindu system of writing down numbers. His book was used for nearly two centuries as a kind of commercial arithmetic by merchants who were not slow to appreciate the advantages accruing from the various methods of accounting and bookkeeping which he expounded.

At a time of quickening interest in arithmetic and algebra it is to be expected that we should find interest in counting extending itself to dice. There is no mention yet of chance (or likelihood), but M. G. Kendall draws our attention to Bishop Wibold of Cambrai, who, some time about A.D. 960, so far forgot the pagan origins of dice and the fulminations of his superiors as to invent a moral dice-game. He enumerated fifty-six virtues—it is surprising to learn there are so many—and assigned one to each

—◄06:0:90►— 19

Quinquaginta modis & fex diverfificantur
In punctaturis, punctaturæque ducentis
Atque bis octo cadendi fchematibus, quibus inter
Compofitos numeros, quibus eft luforibus ufus,
Divifis, prout inter eos funt diftribuendi,
Plenè cognosces, quantæ virtutis eorum
Quilibet esfe poteft, feu quantæ debilitatis:
Quod fubfcripta poteft tibi declarare figura.

Tabula III.

Qvot Punctaturas, et qvot Cadentias ha
beat qvilibet numerorū compofitorum.

3	18	Punctatura	1	Cadentia	1
4	17	Punctatura	1	Cadentiæ	3
5	16	Punctaturæ	2	Cadentiæ	6
6	15	Punctaturæ	3	Cadentiæ	10
7	14	Punctaturæ	4	Cadentiæ	15
8	13	Punctaturæ	5	Cadentiæ	21
9	12	Punctaturæ	6	Cadentiæ	25
10	11	Punctaturæ	6	Cadentiæ	27

Fig. 5—Printed version of part of "De Vetula" (see Plates 6 and 7) from the edition published at Wolfenbüttel, 1662
(By courtesy of Dr. M. G. Kendall and the Editors of "Biometrika")

3-partition of 18, 17, 16, . . . , 3, i.e. one to each number of the different ways in which 3 dice can fall. The dice were cast, the virtue determined, and the thrower then concentrated on this virtue for a certain length of time.* Thus the number of 3-partitions is correctly enumerated, and the fact that the order of the fall of the dice doesn't seem to matter is in agreement with what we have surmised to be the classical oracular tradition. The pagan influence was not easily eradicated.

The number of ways in which three dice can fall, allowing for permutations, is given in Latin verse in a poem entitled *De Vetula* and ascribed by some to Richard de Fournival (A.D. 1200–1250) who was Chancellor of Amiens Cathedral (see Plates 6 and 7, between pages 24 and 25, and Fig. 5). Various other authors have been suggested at varying dates up to the end of the fifteenth century. The writing of the numbers themselves in this document is interesting. In a fourteenth-century German missal the digits are written

$$1 \quad 2 \quad 3 \quad \mathcal{P} \quad \mathcal{I} \quad 6 \quad \wedge \quad \mathcal{S} \quad 9$$

This is the symbolism used here except for the 5, which is like the symbol ascribed to Boethius (A.D. 480–525) who wrote it ς. The partitional falls of the dice are written down in diagrammatic form, probably by a later commentator. There are 56 such cases. The relevant passage in the text of the poem may be translated as follows :

> If all three numbers are alike there are six possibilities ; if two are alike and the other different there are 30 cases, because the pair can be chosen in six ways, and the other in five ; and if all three are different there are 20 ways, because 30 times 4 is 120 but each possibility arises in 6 ways. There are 56 possibilities. But if all three are alike there is only one way for each number ; if two are alike and one different there are three ways ; and if all are different there are six ways. The accompanying figure shows the various ways.

It can be noticed that the commentator, in spite of his correct diagram, gives the wrong number of partitions for 10 and 11 but

---

* One may be pardoned perhaps for reflecting that a pin would have been quicker and less open to the threat of correction by one's superiors.

still achieves the right number of 27 for the number of combinations.*

The idea of combinations, the choosing of a number of things from a group, is not new at this time. It has been noted briefly that the Greeks attempted such calculations, and the Arabs and the Chinese also indulged in them. Omar Khayyam (died *c.* A.D. 1214), poet and algebraist of quality, had some idea of the relation between binomial coefficients, and Chu Shih-chieh in the " Precious Mirror of the Four Elements " (Ssŭ-yüan yü-chien), written in 1303, speaks of an ancient method which he says is not his own, for determining one from another. He gives the following diagram :

```
                            1
                        1 [   ] 1
                    1 [   2   ] 1
                1  3  [       ] 3  1
            1  4  [       6       ] 4  1
        1  5  [   10  ]       [   10  ]  5   1
    1  6  [   15  ]       20      [   15  ]  6   1
  1  7  [       21  35          35  ] 21  7   1
1  8  28  [   56  ]       70      [       56  28  8  1
```

Sarton (q.v.) remarks that " this is an excellent example of recondite knowledge that was many times discovered, lost and rediscovered, until the matter was fully understood". But while the significance of the combinatorial coefficients in relation to the expansion of the positive binomial may have taken some time to be recognised, this does not mean that mathematicians were not familiar with their combinatory uses, and the commentator of *De Vetula* quite obviously was.

The gambling game of hazard† is one of the first dice-games

---

* A very minor point, 16 instead of 18 is written for the number having the same number of permutations as 3. This is just a slip in writing. (My comments refer to the original manuscript reproduction in Plates 6 and 7, and not to Wolfenbüttel's correct version.)

† According to Larousse, from *El-Azar*, a chateau in Syria. According to others from *al-zhar*, the Arabic name for a die.

to be mentioned in the literature of the Christian era. It is played with two or three dice* and is possibly a direct descendent of the oracular mysteries. It has been suggested that the game was brought to Europe by the knights returning from a crusade. This is a possibility but it does not rule out the idea that the game is a debasement of religious rites of classical times. Dante (1265–1321) makes a brief comment on the game in the sixth canto of the *Purgatorio* (*Divina Commedia*) and a later commentator (1477) on Dante has this to say :

> Concerning the number of throws it is to be observed that the dice are square and every face turns up, so that a number which can appear in more ways must occur more frequently, as in the following example : with three dice three is the smallest number which can be thrown, and that only when three aces turn up : four can only happen in one way, namely as two and two aces.

There is, however, no discussion of the number of combinations (as opposed to partitions) which is the key to the problem, although it will be noticed that the idealisation of the die and the concept of equiprobable throws is almost achieved.

The difference in spirit between classical times and the end of the Dark Ages is remarkable. With the Greeks there is on the whole little speculation about the natural phenomena of the earth (as opposed to the heavens) which is based on empirical hypotheses. With the renaissance of the human intellect there is the beginning of the empirical tradition, from observation to hypothesis and back to observation again, which has led to the great strides of modern science. The narrow streets of the Quartier Latin of the eleventh and twelfth centuries were crowded with young men disputing in the manner of Aristotle. Somehow, painfully and slowly, the realisation was born that it is important to see for oneself, a realisation which was fully fledged by the time that Newton proudly stated *non fingo hypotheses*—untrue, but indicative of the spirit of science at that time.

---

* Two dice were commonly used for the game in England in the seventeenth century. (See remarks in Chapter 14.) But there is a strong suggestion that in the early Middle Ages three dice were used in Italy.

If it is possible that the Greeks did not develop the mathematics of die-throwing because of their inability to link up theoretical concept with empirical fact, this may be enough in itself to explain why the probability calculus was so long developing. There are the first few fumblings, which I have noted, at the beginning of the present millennium, but it is possibly no accident that until the influence of the incomparable Leonardo da Vinci (1452–1519) made itself felt with his contemporaries the real search did not begin. So much has been written about this giant of the human intellect that there can be little left to say. All speak of his attitude of mind with its almost reckless desire to know, and to know through observations made by himself, which influenced a turning towards experiment in both the arts and the sciences. He was friendly with two of the earliest protagonists of probability, Fra Luca Paccioli (see Plate 8, facing page 25) and Geronimo Cardano.

Fra Luca Paccioli (or Paciolo), who took the name Luca di Borgo on entering the Franciscan order, was born in Tuscany at Borgo in 1445. Not a great deal is known in detail about his life beyond occasional references in the writings of himself and others. He travelled in the East and taught mathematics at Pérouse, Rome, Naples, Pisa and Venice before he became a professor of mathematics at Milan with some sort of position at the court of Ludovico Sforza il Moro (Lewis the Moor). It was here that he became acquainted with Leonardo who arrived in Milan in 1482. It is related that Leonardo spent his evenings telling stories to the ladies of the court and joining in their pastimes and games.* He was friendly enough with Fra Luca to draw the illustrations for his *De Divina Proportione* (Venice 1509) and, some say, to contribute to the text. Both Leonardo and Fra Luca left Milan with the arrival of Louis XII, Luca to Firenze and Leonardo shortly to Mantua and then to Firenze. Fra Luca died there in

---

* Giovanni Boccaccio (1313-75) describes the recreations of ladies at ducal courts as including both chess and dice. Probably the diversions at the court of the Sforza were similar. No fragment exists to tell us whether Leonardo speculated about the fall of the dice or the combinatorial aspect of chess, but it would have been in keeping with his restless intellect had he done so.

1509. The work for which he is famous from the point of view of the probability calculus is *Summa de arithmetica, geometria, proportioni e proportionalità*, printed in Venice in 1494. There is no doubt that many, if not all, of the mathematicians and scientists of this era were plagiarists, the great exception being Leonardo who seems to have read little and speculated much. This plagiarism was helped by the invention of printing (*c.* 1450), by the great flood of manuscripts released by the sacking of Constantinople, and by the various incursions of the French into Italy. It was apparently legitimate to reproduce an early manuscript with one's own emendations and improvements and to pass off the whole as one's own work.*

Fra Luca incorporated almost in its entirety the *Liber Abaci* of Leonardo the Pisan. The *Summa* is in two parts, one for algebra (arte maggiore) and arithmetic (arte minore), and one for geometry. The arithmetic was for merchants and among other things explained the mechanics of double entry book-keeping. There are many arithmetical examples, and multiplication tables up to 60 × 60 are given. Instead of the eight fundamental operations common at the time he reduces them to seven " in reverence to the gifts of the Holy Ghost". On the whole there is little that is new. Fra Luca was not a great mathematician and his importance lies in the fact that he summarised the mathematical learning which was then common. Perhaps his chief claim to recognition lies in the attempt to combine algebra and geometry in the *De Divina Proportione*, which set the fashion for over a century. From the probability point of view Luca is interesting for one of his examples† in the *Summa*. " *A* and *B* are playing a fair game of *balla*. They agree to continue until one has won six rounds. The game actually stops when *A* has won five and *B* three. How should the stakes be divided?" We would say 7 : 1, but Paccioli argues 5 : 3.

---

* At least that is what it looks like from this vantage-point in history. Possibly one is being a little too censorious, and the fact that much of an author's writings was not original was quite understood by his contemporaries. But see Cardano.

† I have been unable to trace from where he obtained this.

I think it is doubtful whether he realised that the problem
he was discussing was any different from a problem in proportions. However, it is from this work that the impetus came to
Tartaglia and to Geronimo Cardano and for that reason only
(if for no other) it is a landmark ; for the *Summa* of 1494 undoubtedly inspired both Cardano and Tartaglia to write their
versions of the algebra and the arithmetic of the day. Cardano
did not, it would seem, appreciate that there was anything to
do with chance in Paccioli's game of ball. Tartaglia in the
Generale Trattato of 1556 discusses the problem and gets it wrong—
still, in the writer's opinion, thinking of it as an exercise in proportions. I shall return later to the battles between these two
mathematicians. It would seem appropriate here to mention one
further person who discussed this problem. After the publication
of Paccioli's book, the propositions and exercises must have been
the subject of much mathematical disputation ; in fact it seems
almost curious that someone didn't reach the right answer to this
" problem of points " by accident. In 1558, the year after
Tartaglia's death and inspired possibly by the *Generale Trattato*
or by Paccioli's *Summa*, G. F. Peverone published *Due Brevi e Facile
Trattati, il Primo d'Arithmetica, l'Altro di Geometria*. In the arithmetic we find again the poser of *A* and *B* playing for 10 wins, *A*
having won 7 and *B* having won 9. (Still the case of one having 3
to win and the other one only.) Peverone argues :

> *A* should take 2 crowns and *B* 12 crowns. For if *A*, like *B*, had
> one game to go, each would put 2 crowns. If *A* had two games to go
> against *B*'s one, he should put 6 crowns against *B*'s two, because by
> winning two games he would have won four crowns but with the
> risk of losing the second after winning the first. And with three
> games to go he should put 12 crowns because the difficulty and the
> risk are doubled.

M. G. Kendall, who, rightly from some aspects, calls this one of
the near misses of history, points out that if Peverone had stuck to
his own rules, he would have got the right answer. If *B* has one
game to go and is staking 2 crowns, then for *A*,

> with one game to go the stake is 2
> with two games to go the stake is $2 + 4 = 6$

with three games to go the stake is $2 + 4 + 8 = 14$.

However he didn't stick to his rules and so misses the correct solution. Looked at from one point of view, it might be said that here he is trying to enumerate a probability set, but somehow the doubt remains as to whether he really knew what the problem was that he was trying to solve.

References

It will be obvious that in addition to the actual works cited, which are available in many university libraries, I have leaned heavily on

M. G. KENDALL, " Studies in the History of Probability and Statistics : II," *Biometrika* (**43**).

The development of the symbolism in which we write our digits today has been told often. The most comprehensive account, with also a thorough discussion of methods of calculation used in early times, is perhaps that found in the first 250 pages of

D. E. SMITH, *History of Mathematics*, Vol. 1.

The same sort of material is also found in

S. CONNINGTON, *The Story of Arithmetic*

and doubtless in many other places. For general and accurate information I have used Sarton's great learning and in particular

G. SARTON, *Introduction to the History of Science.*

His printed collection of lectures,

G. SARTON, *Six Wings—Men of Science in the Renaissance*

is both interesting and informative.

chapter 5

Tartaglia and Cardano

La distance n'y fait rien ; il n'y a que le premier pas qui coûte.

> LA MARQUISE DU DEFFAND in a letter to d'Alembert, commenting on the legend that St. Denis walked two leagues carrying his head in his hand—*Oxford Dictionary of Quotations*, 170 : 1.

In 1482, as I have noted, Leonardo left Firenze to enter the service of Ludovico Sforza, Duke of Milan. His duties do not seem to have been onerous and for the next seven years he was his versatile and ingenious self in many fields of the arts and sciences. Once in Milan it was only a matter of time before he visited the University of Pavia—a few miles away—for it was there, housed in magnificent splendour, that then existed the fabulous library of Petrarch (1304–1374), bought by Gian Galleazo Visconti and presented by him to the University. Scholars travelled from all over Europe to visit this library, and although Leonardo was no seeker after other men's truths he would almost certainly have paid visits to it and to the University.

Some time early after his arrival in Milan he made the acquaintance of Facio Cardano (1444–1524) who practised as a lawyer and as a physician and who lectured in the University of Pavia in geometry, said to be his favourite subject. Several of Leonardo's biographers speak of his being fascinated by strange characters and, while Facio cannot hope to compete with the record of his son, enough has been written of his pale eyes which shone luminous like a cat's in the dark, of his trembling hands, of his gnome-like figure dressed in scarlet instead of the orthodox black of the professoriate, to explain the beginning of the acquaintance with Leonardo. This acquaintance soon ripened into friendship with a common interest in perspective and optics. Facio was at this time " emending " or " improving " the work of John Peckham.* There have been differences of opinion as to whether Facio's " improvements " were ever published, but Corte, writing in 1718, says that they were published under the title *Prospectiva communis D. Johannis, archiepiscopi Cantuariensis, F. ordinis Minorum, ad unguem castigata per eximium artium et medicinae et juris utriusque doctorem ac mathematicum peritissimum D. Facium Cardanum, Mediolanensem, in venerabili collegio Jurisperitorum Mediolani residentem,†* and Morley, writing in 1854, says that it was printed *c.* 1480 in Milan. It has also been stated that Leonardo's note-books, which were not published until the eighteenth century, served as a mine for Facio's " improvements " to the *Prospectiva.* Whether any of this is true or not, it seems undeniable that Facio was talented but not creative and was possibly not liked by many, although the opprobrium which was the lot of his son may have been transferred to him by later writers. His contemporaries did, however, we are told, look on him as a miracle of learning and he is said to have lived a hard and strenuous life, showing no consideration to himself or to those who lived or worked with him.

In fifteenth-century Italy illegitimacy was no bar either socially or professionally, and it was usual for children born out of wedlock to be brought up with and receive the same advantages as those born in it. Leonardo himself was illegitimate, the son of a lawyer and a peasant girl, and so was the first known son of Facio Cardano. To this eccentric and to Chiara Micheria was born at Pavia on September 24th, 1501, an illegitimate son who was baptized Girolamo. Facio is said to have been descended from a family of lawyers and doctors, but little is known about that of Chiara. She is spoken of as being a short, fat, healthy

---

* John Peckham—a Franciscan Friar who became Archbishop of Canterbury in 1279 on the nomination of Pope Nicholas III.

† " . . . the excellent doctor in the arts as well of medicine as law, and most experienced mathematician, Facio Cardano of Milan, residing in the venerable college of the Milanese jurisconsuls."

woman, young and with a lively wit. From descriptions of her later in life it has been thought that there was some evidence that she suffered from epilepsy, although this is not certain. What is certain is that in her son, Girolamo, she had given birth to one of the greatest eccentrics of all times. Knowledge of this son might have been equally fragmentary had he not written his auto-biography *De Vita Propria*, a writing which all his biographers agree is unique, and which for needless frankness, self-pity and invention leaves far behind even the famous *Confessions* of Jean-Jacques Rousseau. The interesting thing about *De Vita Propria* is that although it was written at the end of a long life, when the old man was allowed by all who write about him to be suffering at least from an ardent persecution mania, yet the statements in it about his achievements have been often readily accepted. This seems odd, partly because where factual checks are possible they show that on occasion he was not without the power of grandi-loquent invention, and partly because, the autobiography having been written for posthumous publication, it was not suspected that he was writing to interest future generations in himself rather than to tell the truth.

He says about himself :

> I was ever hot-tempered, single-minded and given to women. From these cardinal tendencies there proceeded truculence of temper, wrangling, obstinacy, rudeness of carriage, anger, and an inordinate desire for revenge in respect of any wrong done to me. I am moreover truthful, mindful of benefits wrought to me, a lover of justice, a despiser of money, a worshipper of that fame which defies death, prone to thrust aside what is commonplace and still more disposed to treat mere trifles in the same way.

He tells us that he is able to understand languages without learning them ; that one day he bought a copy of Apuleius and he found the next day he could read it easily without previously having opened a Latin book. Again :

> Whenever I have incurred a loss I have never been content merely to retrieve the same ; I have always contrived to seize something extra,

which may not be entirely consistent with despising money.

Cardano's interest from the probabilist's point of view lies in the fact that he wrote *Liber de Ludo Aleae* (Book on games of chance). This was found among his manuscripts after his death and was not printed until 1663 in Basle. We are told that it was actually written in 1525 and rewritten in 1565, but the evidence for this rests entirely on Cardano himself. It does appear likely, however, that some time during the 74 years of his troubled life the idea of the equi-likely points of the fundamental probability set was born, but when and how can only rest on conjecture. In order to try to arrive at what is possible it is necessary to contemplate Cardano's writings and his relationships with other mathematicians of this sixteenth century, to search for understanding by considering the pattern of his life. The difficulties loom large, for all the mathematicians were vain, boastful, aggressive and by the standards of today, plagiarists, so that Cardano's abnormalities needs must be judged against an abnormal background.

For the first few years of Cardano's life his mother and father did not live under the same roof. They appear to have joined up when Girolamo was about five years old, and shortly after this his father pressed him into service to carry his bag when he went to pay his visits about the city. For some reason Girolamo thought this an indignity, although Facio appears on the whole to have been a fond father. He instructed his son in mathematics and was perhaps opposed to his reading medicine. Girolamo writes that his father wanted him to study law, but since the boy had had no formal instruction in Latin this is possibly doubtful. Girolamo's early interests were undoubtedly in medicine and in 1520 he enrolled at the University of Pavia as a medical student, being already noted, according to his own account, for excesses at the gaming tables. For the next three years he spent his time between Pavia and Milan and at the end of this time had learnt enough to teach Euclid, Dialectics, and Philosophy. In 1523 the University of Pavia closed, temporarily, through lack of funds, and Cardano, still interested in medicine and the mathematical problems of the day, went first to Milan and then to the University of Padua. Writing fifty years later he says he was elected

Rector of the Academy of Padua at the close of his twenty-fourth year (1525). Waters, one of his biographers, points out that this is unlikely. By tradition Pavia is one of the seats of learning founded by Charlemagne. Padua is not so old, having been started through a migration of students from Bologna in 1222. Nevertheless a man of some social standing was usually elected as its Rector. The University was closed from 1509 to 1515 because of the unsettledness of the country caused by the invasions of the French armies, and the first Rector after this period was not elected until 1527. The records at Padua are still extant and cover this period, but there is no mention of Cardano's name in the list of Rectors.

It may be that all Cardano meant was that he held some student office in the University, or he may have invented his post to excuse himself to posterity for having quickly spent the small patrimony which came to him about this time on the death of his father. Whatever his motives there is no doubt that this was a time of great extravagance by him. In 1526 he was elected a Doctor of Medicine of the University, having been rejected twice before the third successful ballot. During this time at Padua, Cardano says he supplemented his small income by gambling. He probably did gamble but that he did so successfully does not entirely accord with the known fact that he ran through his inherited money.

In 1527, when he was in his twenty-sixth year, he began to practise as a physician in the small village of Sacco, and in 1529, for the first time, he attempted to gain admission to the College of Physicians of Milan. This admission was refused, ostensibly on the grounds of his illegitimate birth, but, since this was a rule more honoured in the breach than in the observance, it was possibly because of his reputation as a gambler and his licentious conduct while at Padua that the physicians did not deem him suitable for a colleague. The village at Sacco is said to have been a poor one and Cardano did not fare very well. In 1531 he met and married Lucia Bandarini, moving to Milan for a few months in 1532. While in Milan he made another unsuccessful attempt to enter the College of Physicians.

The tenacity of purpose which he showed in his attempts to be elected to this college is to be noted since it appears again in other situations and is obviously one of the cornerstones of his character. Each time it is met it comes as a surprise. In this particular connection it is a little difficult to see what Cardano thought he would gain commensurate with the effort which he put out. His doctorate from the University of Padua enabled him to practise medicine, and one can only suggest that possibly he thought the added respectability of the membership of the college of Milan would increase the number of his patients. But whatever his motive, this time again the membership was not to be and he took himself and his family off to Gallerate, another poor village where again he did not fare very well. He spent his free time gambling assiduously and reading voraciously anything he could lay his hands on. Possibly the former distraction was the reason why he returned to Milan in 1534 and established himself with his wife and child in the poorhouse.

At this point in his variegated career he had a run of good luck. A friend obtained for him a lectureship in geometry at the University of Milan, and the moneys from this lectureship, although small, were enough to enable him and his family to live. Also at this point his mother decided to forgive him for his extravagances at Padua and came to live with him. With his mother's small income and his own earnings the Cardano family were able to exist in a comparatively good style. Cardano, however, seems to have been one of those persons born without the instinct of self-preservation. In 1536 he produced his first book, *De Malo Recentiorum Medicorum Medendi Usu* (On the bad practice of medicine in common use), in which he was concerned to point out what he considered to be the faults in the prevailing accepted treatment of illnesses. (Hardly the way, one would think, to endear himself to the Milanese physicians.) He was found to be inaccurate in many of his statements and was humiliated in the ensuing quarrelling. But 1536 was to be a year of fate for him in yet another way.

"On the day when Ludovico Ferrari came from Bologna with his cousin Luca, the last day of November 1536, the magpie in the courtyard kept up such an endless and altogether unwonted chattering that we were looking for someone to arrive." Luca had been a servant in Cardano's house for some months and on his return from Bologna brought his cousin Ludovico (1522–1565) with him, and he also entered Cardano's service. Most if not all that we know about Ludovico as a person is what Cardano recounted in the memoir which he wrote about him, a source which in this, as in many other contexts, is suspect. Ferrari is said to have been little, neat and rosy, with an agreeably modulated voice, possessing a cheerful face with a snub nose. He was reputed to be fond of pleasure, openly irreligious and habitually scornful of those who were not. His temper was hot and uncontrolled so that it was necessary to watch one's words when talking to him. In spite of his extraordinary mental ability he was quite uneducated when he entered Cardano's service at the age of fifteen.*

Cardano, having been appointed a teacher of geometry (which also included arithmetic, algebra, astronomy and astrology) and having been savaged at the hands of the medicals, might have been expected to turn to other fields of endeavour and this, in part, he did. He perhaps wanted to recoup his somewhat shattered reputation, and possibly also he was again lacking money, for he gambled so strenuously (and presumably unsuccessfully) that he was short of money and goods at many periods of his life. It seems, however, more than fortuitous that this, the mathematical period of Cardano's life, coincided with the presence in his house of Ludovico Ferrari. Cardano instructed the boy and used him as an archivist and as a copyist of manuscripts. He was repaid by a passionate devotion, which apparently lasted for the whole of Ferrari's short life, so that he was accustomed to speak of himself as "Cardano's creation".

In 1539 the first of Cardano's mathematical books was published in Venice and the author received ten gold crowns for it.

---

* Some biographers state that he was eleven when he became Cardano's servant. The dates of his life (1522-65) are those commonly accepted, and the date 1536 is that given by Cardano, which would make him fourteen or fifteen. An appreciation of the life and work of Ferrari is long overdue.

Entitled *Practica Arithmetica et Mensurandi Singularis* (Practical arithmetic and simple mensuration), it was based largely on Paccioli's book of 1494.* In an appendix to his own book Cardano is concerned to point out the errors in this work but does not trouble to say that he derived his book from it. He was in fact " improving " and not originating. Later in correspondence with Tartaglia he makes the statement that he has done nothing but correct Fra Luca's errors, although the statement in this context is suspicious. One may imagine Cardano instructing Ferrari with the aid of Paccioli's book and the keen young mathematical intelligence being brought up short by the gaps in the reasoning. Cardano's book gave him a certain mathematical reputation. He would also know that any further treatises of this type were saleable. He did not cease from his efforts to get recognition from the Milanese physicians, and after one more unsuccessful attempt in 1537 finally achieved his desire in 1539.

I will now turn briefly to the famous quarrel about the solution of the cubic equation. Although this does not have relevance to the development of probability concepts it is important in that it illustrates the sort of person Cardano was. The whole episode throws into relief once again the pertinacity of purpose which Cardano had already demonstrated in obtaining entrance to the College of Physicians at Milan, and, to the present writer at any rate, shows his reliance on the mathematical powers of Ludovico.† Leonardo the Pisan had solved equations of the first degree, and Fra Luca Paccioli went so far as to solve quadratic equations which had positive roots. He further declared that to look for solutions of equations of the third and fourth degree was as hopeless as trying to quadrature the circle, a remark which put off some mathematicians from attempting the impossible

---

* I have studied Cardano's book to see if the game of *balla* is discussed but cannot find it. However I may have missed it. It would be very interesting historically if Cardano had discussed it and got it wrong (or right!).

† The complete story of the " cubic " quarrel is told in many histories. See for example J. F. Scott's *History of Mathematics*, pp. 87 *et seq.*, for mathematical discussion.

but which only stimulated others to try.*   In 1536 a Brescian schoolmaster-mathematician called Giovanni Colla visited Cardano and told him that Scipio Ferreo (d. 1526) of Bologna had discovered two new rules in algebra for the solution of problems dealing with cubes and numbers. (Actually it was the rule for the solution of one case of a compound cubic equation.) Colla said that Nicolo Fontana (1500–57), better known as Tartaglia—the stutterer—and Antonio Maria Fiore also knew these rules, Fiore having been told them by Ferreo, his teacher. It was generally accepted that Tartaglia invented them for himself. One of the ways in which a mathematician obtained a reputation in those days was to go to his own or another university and challenge any one publicly to solve a particular problem. Fiore had had a successful time challenging mathematicians to solve equations of the type

$$x^3 + px - q = 0.$$

History does not relate how he solved them, but his, and his master's, reputation stood high when he disputed publicly with Tartaglia at Venice in 1535. For the contest it was laid down that each contestant should ask the other thirty questions and Fiore should begin. Tartaglia answered all Fiore's problems correctly and then he proceeded to ask Fiore thirty questions involving a knowledge of the solution of

$$x^3 + px^2 - q = 0 \quad \text{and} \quad x^3 - px^2 - q = 0,$$

only one of which was Fiore able to answer. Great was the glory of Tartaglia and even greater was the desire of Cardano, when he heard about the contest, to know how he did it.

The anxiety of Cardano to know is one of the more puzzling aspects of this episode. He pestered Tartaglia, knowing that Tartaglia was only supreme as long as he kept his secret. It was

---

* To illustrate the difficulties of the early algebraists we take an example from Paccioli quoted by Scott (q.v.). A multiplication sum is written as follows :—

    4.p.R6 4.m.R6   Productum   16.m.6   10.

or, as might now be written,

$$(4 + \sqrt{6})(4 - \sqrt{6}) = 16 - 6 = 10.$$

We do not realise our own good fortune.

therefore unlikely that Tartaglia would have told anyone even if he had been a gentle kindly man, and from descriptions of him he was far from being this : he is depicted as being churlish, uncouth and ill-mannered, rendered suspicious by his very humble birth and extreme poverty of anyone trying to take advantage of him. But Cardano, once determined on this course of action, showed his characteristic tenacity of purpose. There were several exchanges of letters between him and Tartaglia in the years after 1536 in which the acerbity with which each wrote becomes more marked with time, and it would seem from these exchanges that Cardano had tried to solve Fiore's problems and failed. In 1537 Tartaglia paid a visit to Milan, and Cardano, whether by trickery (as is said by some) or because Tartaglia felt somewhat contemptuous of his mathematical powers (probably this latter), succeeded in extracting from him under secrecy a riddle which he said embodied his procedure. To judge from the enquiries with which Cardano pursued the Brescian later, the riddle remained unresolved and the enquiries—needless to relate— remained unanswered.

Ferrari was, by 1540, showing intellectual maturity, and although only eighteen was lecturing in mathematics when Colla again visited Cardano. He asked for a solution of

$$x^4 + 6x^2 + 36 = 60x$$

which it is supposed he found in some Arab treatise. Since both he and Tartaglia lived at Brescia it is possible that he had already visited Tartaglia with the problem, but history does not relate this. Cardano was unable to solve this equation using what he thought to be Tartaglia's methods, but Ferrari worked on the problem and eventually produced a rule for the solution of biquadratic equations from rules for the solution of the cubic. Cardano, if he had been the honest man he thought himself to be, was now on the horns of a dilemma. Ferrari had made a great discovery. It could be said by his contemporaries that the discovery had only been made because Tartaglia had disclosed his rules under secrecy, and Tartaglia still showed no signs of wanting to disclose them publicly. Cardano appears to have hesitated for some years, but being the man he was, greed for money and perhaps greed for fame were too much for him. He published an algebra, *Artis Magnae Liber* (the Book of the Great Art) in 1545.

Cardano was a man about whom it is difficult to preserve a judicial equilibrium. Some of his biographers write as though he was a persecuted saint, others as though he was an unpleasant charlatan. But however one may explain away the numerous doubtful episodes of his rather sordid life, the correspondence with Tartaglia shows him to be at this time both a liar and dishonest in his actions. Further if it is remembered that it was Ferrari who produced the rules after Cardano had stumbled over Tartaglia's hints, the piquancy of his remarks on the subject of algebra are still to be relished.

> And even if I were to claim this art as my own invention, I should perhaps be speaking only the truth,* though Ptolemy, Paccioli and Boëtius have written much thereon. For men like these never came near to discover one-hundredth part of the things discovered by me. But with regard to this matter—as with divers others—I leave judgment to be given by those who shall come after me. Nevertheless I am constrained to call this work of mine a perfect one, seeing that it well-nigh transcends the bounds of human perception.

In the long and bitter wrangling which took place after the publication of the *Artis Magnae Liber* Cardano took no part. Ferrari and Tartaglia exchanged animosities, finally meeting in a public disputation in which each claimed to have come off the better but in which Ferrari appears to have routed his opponent. The fact that Cardano took no part in it, in spite of all Tartaglia could do to bring him into the arena, is probably indicative of the fact that he recognised his own limitations and did not want to be exposed by one who was, by now, his implacable enemy. Ferrari, calling himself " Cardano's creation " and battling vigorously against Tartaglia, shows himself far superior to either of them, and his loyalty shines down the years.

In the *Artis Magnae Liber* is the first suggestion that the square of a number may have positive or negative roots, that the roots

---

* " What is truth ? " said jesting Pilate, and would not stay for an answer.

of an equation cannot always be found, and the trick of changing a cubic equation into one in which the squared term is missing. Cardano says that Ferrari only invented two of his rules, that he got three others from the algebra of Mahommed ben Musa, and that he invented all the rest himself. Montucla, writing in 1758, and who many of Cardano's biographers assert has a strong bias against Cardano, will only allow him the credit of the negative and positive roots. It is possible that little was his, it is possible that he did much, but whichever it was, the whole story bears out the hypothesis that Cardano was an improver rather than an originator, and that he was not a mathematician of the class of Tartaglia or Ferrari. One may note with some interest that certain medical historians would hold that his contributions to medicine have not stood the test of time and that his fame rests principally on his contributions to mathematics.

Two years after his disputation with Tartaglia, Ferrari seems to have passed out of close contact with Cardano. He held for a short time the office of Surveyor of the province under the Governor of Milan and then retired through ill-health, perhaps brought about by dissipation, to live with his sister at Bologna. He was appointed to a Professorship of Mathematics there in 1565, the year of his death. And with the departure of Ferrari from his immediate circle ends Cardano's brief excursus into mathematical research. In *De Subtilitate Rerum*, a collection of essays containing a mixture in about equal parts of the natural sciences, folklore and superstition, there is a suggestion of a system for writing down numbers, but he can hardly have expected this to be taken seriously. The rest of his life we shall recount briefly before turning to the book on games of chance.

Cardano remained at Milan for some years, lecturing in geometry at the University at Pavia, practising the profession of medicine, and writing books on medicine and the natural sciences. These books, much of which are nonsensical by modern standards, were thought highly of at the time of their writing. He was a voracious reader, and while it was accepted that his books contained nothing new, they were a synthesis of the old and the then modern learning and summarised the scientific knowledge of the day. These were Cardano's golden years. It is true that his wife Lucia died in 1546, but this does not seem to have been a mortal blow. The gossip about his predilection for singing boys which was whispered when he was a student revived at this time, but this was a favourite smear of the age ; Leonardo had to suffer the same imputation. Cardano was making money and beside his mathematical fame had an enormous reputation in medicine, so much so that he was called to St. Andrews in 1552 to treat John Hamilton, the Archbishop. He did this successfully, but was not so successful in his passage through London when he cast the horoscope of Edward VI and forecast a long life for him ; the king died the next year. However from 1547 to 1557 he had everything that he could wish for—fame, position, money, respect. Then the Parcae, the accursed sisters, struck devastatingly. Cardano's eldest son, Giambattista (1534–1560), married a prostitute of Milan in 1557 and in 1560 poisoned her with white arsenic and was executed for her murder.

Cardano showed his tremendous tenacity of purpose yet again in his fight for the life of his son, but was unsuccessful and his failure seems to have driven him mad. Throughout the rest of his life he seems to take the attitude that he is the target of attack by the Fates, but that those whom the Fates choose as their instrument to work against him die quickly. There is continued repetition in his writing that someone or other did him some sort of injury—probably imaginary—and then a note of glee about their death shortly afterwards. Throughout this disappointing time he still keeps up the habit of work. After his son's death he reported he was writing two treatises, one on dialectics and one on anatomy. But men follow their own characters all through life and, as we have previously remarked, he seems to have had no instinct for self-preservation. Cardano's lectures are said to have been deserted, either because there was truth in the statement that they were incoherent and inconsequent or because his son's execution as a convicted felon made the students avoid him. Yet at such a time, when one would have thought it prudent to walk carefully, his conduct was such that the University was moved to protest about the number and character of the young men who frequented his house. This may have been the University's way of getting rid of him, since he himself vigorously protested his innocence. However, this last is a little doubtful since his daughter wrote him a letter saying that she was ashamed of her kinship with him. His persecution mania appeared to have intensified, so that all protests, however kindly, were put down to malignant intent, and he chronicled many attempts to kill him both by poison and by dropping heavy objects on his head. Forced to resign his professorship at Pavia in 1562 he lived in Milan for a short time until, through the intercession of Ferrari, he was appointed to a professorship of medicine at Bologna. His licence to teach in Milan had been revoked, presumably as a consequence of his conduct at Pavia.

The immortals had not yet finished their sport with the old man. Ferrari, appointed to a professorship of mathematics at Bologna in 1565, was poisoned, it is said, with white arsenic by his sister in the same year. This seems to have driven Cardano still further into madness. By all accounts he spent his time quietly enough revising his old manuscripts and perhaps adding to them, but it is possible that the quality of his teaching had deteriorated and it may be that in 1570 the University of Bologna had had enough. They had treated him kindly and with deference, but it is difficult to see otherwise why he was arrested on a charge of impiety because many years (1539) before he had cast the horoscope of Jesus Christ ; for it was not unusual for astrologers to do this. He was placed under house arrest but released after a short time in order to travel to Rome with Rudolfo Sylvestro, one of his pupils who was going there to practise medicine. Sylvestro looked after him until he died in Rome on September 20th, 1576. The uncharitable gossip was that he had predicted he would die on that day and had starved himself for three weeks before to make it come true. At his death he had 131 printed works to his account and 111 books in manuscript. He says that he burnt a further 170 manuscripts which he considered to be without interest.

References

Cardano seems to have been a person who excited biographers' curiosity and many books have been written about him. The fullest details of his life are possibly found in

HENRY MORLEY, *The Life of Girolamo Cardano of Milan, Physician,* with supplementary information in

W. G. WATERS, *Jerome Cardano: a biographical study.*

It seems difficult for biographers to take the " middle road " with Cardano, and Morley and Waters are not, on the whole, favourable to him in the interpretation of the events of his variegated career. If a corrective is desired after reading these, an account of Cardano which presents him as a persecuted savant will be found in

O. ORE, *Cardano: The Gambling Scholar.*

This latter book contains a translation of *Liber de Ludo Aleae* by S. H. Gould. Ore gives some references. The *Nouvelle Biographie Universelle* edited by Hoeffer contains much information in the section in Cardano written by Victor Sardou. Sardou gives references to the writings of de Thou and of Teissier both of which are pertinent in this context, and he seems to have missed very little.

Cardano's autobiography *De Vita Propria Liber* has been translated by Jean Stoner (*Book of my Life*). I would refer again both to the works of G. Sarton and of J. F. Scott, *A History of Mathematics.*

chapter 6

Cardano and Liber de Ludo Aleae

> To throw in a fair game at Hazards only three spots, when
> something great is at stake, or some business is the hazard,
> is a natural occurrence and deserves to be so deemed ; and
> even when they come up the same way for a second time, if
> the throw be repeated. If the third and fourth plays are the
> same, surely there is occasion for suspicion on the part of a
> prudent man.
>
> CARDANO, De Vita Propria Liber

Celio Calcagnini, philosopher, poet and astronomer, was born at
Ferrara on September 17th, 1479 and died in that city on August
27th, 1541. The illegitimate son of a church dignitary, he was
acknowledged by his father and rose to become a canon of the
cathedral at Ferrara and a professor of belles-lettres at the
University. Besides being an astronomer of repute—he wrote
a book to show that the earth revolved around the sun—he was
also interested in the manners and customs of classical times and
before, and wrote several books on this subject. From the point
of view of games of chance it may be noted that he wrote *De
talorum, tessarum ac calculorum ludis ex more veterum*. In this book
there is an historical account of dice and astragali in the classical
era and learned discussion as to how the astragali (for example)
had to fall in order to achieve the venus-throw. There is, however,
no discussion on the calculation of the chances involved, and this
perhaps indicates that mathematical speculation about chance
was then infrequent if not entirely absent. Calcagnini's book may
have been a starting-point for Cardano since he acknowledges
information from it in his own book on chance. Just how Cardano
wrote his books is interesting. He says (*De Subtilitate Rerum*) :

55

> According to my custom I used the following procedure. First I collected a conglomeration of facts about games of chance and then this was later expanded into four books. The first of these was actually completed and dealt with the game of chess : it contained one hundred pages.

This was, one can well believe, typical of the way in which he worked. He read widely, collected a mass of assorted information and then wove it into a whole, adding bits wherever he felt able. The idea of what to write about he probably also got in the same way. His medical books and his autobiography possibly owe their provenance to Galen. The original idea for his books on games of chance and their content would probably have been inspired by Calcagnini. For it seems unlikely in the light of history that in Cardano's original plan the calculation of chances was contemplated.

According to his autobiography, Cardano began to gamble at an early age and he started collecting his material for the book on games of chance while he was at Padua (*c.* 1526). At the back of his Arithmetic, published in 1539, he prints as a kind of advertisement, a consent which the Emperor Charles V has made to a petition from him. Cardano had asked that he might print thirty-four more unpublished manuscripts and that other persons should be prohibited from printing them against his wishes. This consent of the Emperor seems a remarkable *carte blanche*, but it is noteworthy from our point of view in that the book on games of chance is not listed (although the horoscope of Jesus is). Possibly this book on chance was still in the " collecting " stage, for Cardano was in effect advertising his wares, and if it had been anywhere near completion I think he would have mentioned it. Knowing the contents of the book which has come down to us and that he writes of having expanded his original scheme to encompass four books on games of chance, it appears likely that in these early years he was just collecting data for his book on chess. In *De Vita Propria* at the opening of the chapter on dicing and gaming he writes :

> Peradventure in no respect can I be deemed worthy of praise ; for so surely as I was inordinately addicted to the chess-board and

the dicing table, I know that I must rather be considered deserving of the severest censure. I gambled at both for many years, at chess more than forty years and at dice about twenty-five.

De Vita Propria was written about 1574, so it would appear that he did not become interested in dice-games until about 1550. This checks with another remark which he made, that he did not become interested in dice-games until his sons introduced them into his house. The reference to chess as a gambling game is at first sight puzzling since nowadays this game would appear far removed from a gamble. It is suggested that the game was played fast and that considerable money changed hands at various stages, so that anyone who had analysed the possible moves would be at an advantage. Cardano writes :

> Although I have expounded many remarkable facts in a book on the combinations in chess, certain of these combinations escaped me because I was busy with other occupations. Eight or ten plays which I was never able to recapture seemed to outwit human ingenuity and appeared to be stalemates.

In *De Subtilitate* (1551) he wrote that the second book was on games such as dice and primero (about twenty pages), and the third and fourth on games which are a combination of luck and skill. But it is not certain that these books were even then in existence. Possibly at this time they were still in the formative state, for Ore points out that Cardano tells a story in *Liber de Ludo Aleae* about the year 1526 and then says he is a little unclear about details since it happened nearly thirty-eight years before. This would mean that he was writing in the years 1563 or 1564, when, it is interesting to note, he had gone to Bologna and had made contact again with Ferrari. It is possible that the advice to gamblers, the moral reflections and the classical notes, which form part of the book on games of chance, were part of the accumulation of facts which Cardano describes as the way in which he started on a topic. It is also possible that the attempt to link up theory and practice which occurs in this book arose, at least, out of conversations with Ferrari. Further, although this may have been due to the fumbling expression of an essentially new idea, there is more than a suggestion in the work that Cardano did not really know what he was doing.

As M. G. Kendall has pointed out, the intensity of gambling throughout the ages had been such that the relative probabilistic worth of the various hands at cards and of throws of dice had been settled empirically. The zeal and passion which Cardano devoted to the actual playing of games of chance would certainly mean— for he was a not-unintelligent man and an inveterate gambler— that he had a good working knowledge of these empirical chances. Also, since his reading was wide in the process of his collection of facts, it is not fanciful to assume that he knew of *De Vetula* and the several commentaries on it, where the fundamental probability set of partitions is correctly enumerated. The step which he or Ferrari took, and it is a big one, is to introduce the idea of combinations to enumerate all the elements of the fundamental probability set, and to notice that if all the elements of this set are of equal weight, then the ratio of the number of favourable cases to the total number of cases gives a result in accordance with experience.

The crucial chapter " On the cast of one die " in Gould's translation of *Liber de Ludo Aleae*, which is incorporated in Ore's book, is confused : the significant passage runs as follows :

> One-half the total number of faces always represents equality ; thus the chances are equal that a given point will turn up in three throws, for the total circuit is completed in six, or again that one of three given points will turn up in one throw. For example I can as easily throw one, three or five as two, four or six. The wagers therefore are laid in accordance with this equality if the die is honest.

There is no doubt about it : here the abstraction from empiricism to theoretical concept is made and, as far as I know, for the first time. Cardano goes on to say :

> . . . these facts contribute a great deal to understanding but hardly anything to practical play.

Ore interprets this to mean that Cardano was pointing out that most dice-games used more than one die. I think this is not so and that the remark is one made by the gambler as opposed to the mathematician. For to say that a probability is 1/6 is interesting to the mathematician but does not tell the gambler what the outcome of any particular throw is going to be. There are succeeding chapters in the book on the cast of two dice, of three dice, and various combinations connected with these. Given that the elements of the fundamental probability set were correctly enumerated, as they were, it remained for Cardano to calculate the odds, which he did. These calculations are succeeded again by those for card games, the empirical odds for which had already almost certainly been accurately established. Cardano strikes me as being not too happy with these, but he was in the fortunate position of being able to check his theoretical computations from practical experience.

It is when one reads the section on knucklebones that a doubt as to his understanding springs to one's mind. The probabilities are here discussed as though the fundamental set consists of four equally likely elements. It is true that astragali vary so that no two will throw the same, yet it is extremely unlikely that any one of them will have equiprobable sides. This latter is the assumption, however, which he makes in his calculations, and the odds of the various throws are discussed on the supposition that this basic hypothesis is true. The explanation possibly is that Cardano had never played with astragali. He had played with dice, and perhaps noticing empirically that each side of a die turned up on approximately one-sixth of occasions " if the die be honest," he plunged from there and obtained theoretical results for 2 and 3 dice which accorded with practice—the modern scientific " method " in fact. If he had ever gambled with astragali he might have generalised still further, although this is unlikely. The other chapters in the book are made up of advice to gamblers, of why the book was written, of what the ancients had to say about gambling and so on, much of it perhaps taken from Calcagnini. They are interesting for the light that they throw on Cardano, but are irrelevant to the calculus of probabilities.

Posterity will grant Cardano a little more achievement than Todhunter allows him, but it will not, in my opinion, grant him much more. Claims that he anticipated the laws of large numbers cannot really be substantiated. The old man had enough empirical knowledge to be able to judge from experience, but his calculations would not seem to link up with this. It is difficult in these matters not to have too much hindsight and yet to give credit where credit is due. It can be said definitely that as far as is known at present Cardano was the first mathematician to calculate a theoretical probability correctly and that is possibly enough glory. It is interesting to speculate about Ferrari. He is a curious and not unattractive figure of whom we know little except through Cardano. He cared little for reputation and much for money, which considering his humble beginnings is understandable. His loyalty to Cardano as the man who made him appears to have been absolute in an age where treachery was commonplace. His mathematical abilities seem in advance of those of his contemporaries, and yet mathematics to him was apparently just the way in which he could best earn his living. He would probably never again have reached the heights of his solution of the biquadratic, but his early death at the age of 43 leaves one with a feeling of waste. All his MSS disappeared after his death, so that there is no record left of him.

Liber de Ludo Aleae was left in manuscript form at Cardano's death and did not appear in print until the publication of *Hieronymi Cardani Mediolanensis philosophi ac medici celeberrimi opera omnia, cura Car. Sponii*, Lyon, 1663, ten volumes in folio.

chapter 7

Galileo

> To this end was I born, and for this cause came I into the
> world, that I should bear witness unto the truth. . . . Pilate
> saith unto him " What is truth ? "
>
> ST. JOHN, xviii. 38, 39.

During the fifteenth and sixteenth centuries the Hindu method of
number-writing spread through Europe and across the Channel to
England. Also, as might have been expected, algebraic symbolism
improved, the signs plus $(+)$ and minus $(-)$ making their first
appearance. They may be found in the *Arithmetica integra* of
Michael Stifel (1486–1567) which was published in Nuremberg in
1544, but the symbolism may have originated with Christoff
Rudolff and his *Die Coss* of 1525. The English also showed their
first stirrings. Robert Record (1510–58) uses the plus and minus
symbolism in his *Grounde of Artes* published in 1540. Seventeen
years later in 1557 he published the *Whetstone of Witte* in which,
for the first known time, the sign $=$ for equality appears. He chose
it, he endearingly remarks,

> because noe 2 thynges can be more equalle than a pair of paralleles.

It was some few years before this equality sign was universally
adopted.

The arithmetic triangle of the binomial coefficients continued
to appear during this time. It is found on the title-page of the first
edition of *Eyn neue und wolgegründte Underweysung aller Kauffmannsz
Rechnung* (Ingolstadt, 1527) by Petrus Apianus (1495–1552), and
it appears in Stifel's *Arithmetica* up to the seventeenth order, in
De numeris et diversis rationibus (Leipzig, 1545) and in the *Generale*

Trattato (Venice, 1556) of Tartaglia. In no case is any reference made to anyone else, although it would appear too much of a coincidence that all these were original. There is also the post-humous publication of William Buckley in 1567 (see Appendix 1) in which the number of combinations of *n* things taken 2 at a time is fumblingly sought after and found. No surge forward is apparent in the combinatorial field, however. One may note again how much algebraic development, particularly in combinatory analysis, waits on the invention of an adequate symbolism.

Throughout these centuries men continued to gamble with both cards and dice, and somehow among mathematicians the knowledge of how to calculate a probability percolated. It is impossible to say how this knowledge was spread. It appears unlikely that it was known at the time of the *Generale Trattato* of Tartaglia. Cardano either invented it—or at the very least knew of it—possibly about 1563 or 1564, but it is not easy to see how information came from him. He was almost isolated in Bologna and had no mathematical pupils, and Ferrari was dead in 1565. There has been no record yet found of anyone else in the sixteenth century having discussed probabilities, but it is difficult to resist the feeling that somewhere this record must exist, for such calculations were no secret to Galileo-Galilei when he wrote on die-throwing some time between 1613 and 1623. There are various possibilities, all speculative. We may feel that the time was ripe for the discovery to be made. Again and again in the history of mathematical ideas we become convinced of the inevitability of discovery, a feeling which usually arises from a consideration of the false starts, showing that discussion about the particular (or allied) topic was going on in many places. With the probability calculus we have Paccioli, Tartaglia, the near miss of Peverone, the urgency of the gamblers who asked for the calculation of the odds involved, and a revival of an interest in the theory of numbers. It would not therefore be surprising to find several mathematicians writing on chances in gaming in the period 1550–1650. The curious thing is that there was only one Galileo-Galilei, and he obviously did not write as an originator. Either therefore there is a manuscript that we do not know about, and this MS might

equally well have been accessible to Cardano as well as Galileo;
or Ferrari was closely associated with Cardano on the matter
and his pupils from Bologna spread the idea as they wandered
about Italy; or Galileo thought of it for himself. I would be
inclined to rule out the last, since Galileo was at least 49 years of
age before he wrote his fragment and does not appear very
interested in the problem.

Yet another possibility is that gaming problems were discussed
at the meetings of learned societies. It is known that many
societies were founded in Italy during the late sixteenth and early
seventeenth centuries. There is said to have been over seven
hundred altogether in Italy, and although most were in Rome,
the provincial cities also had a number. Firenze is reported to
have had over twenty. Some of these societies or academies were
serious, but usually they seem to have been convivial gatherings
at which papers were read on art or on literature or, occasionally,
on the sciences, but also at which there was gossip and drinking.
John Milton on his visit to Firenze in 1638 was present at some
of these gatherings and read his poems. The possibility does
exist that, through some such social group, ideas about gaming
were disseminated, although this, like all the other possibilities, is
not very probable. What is undeniable is that Galileo produced
a fragment on the subject.

In 1564, the same year as saw the birth of Shakespeare,
Galileo-Galilei was born in the town of Pisa. He was educated
at the Jesuit monastery of Vallombrosa and was so happy at
school that he wanted to stay as a novice.* His father, however,
had other ideas and wanted him to become a merchant. Galileo,
withdrawn from school at the age of 16, showed no aptitude for
commerce, being obsessed with the mechanics of machines in a
way reminiscent of Leonardo who preceded him. He was sent
to the University of Pisa to study medicine, but it was to a Professorship of Mathematics there that he was elected in 1589 (at
the age of 25) through the favour of Ferdinando dei Medici, the
Grand Duke of Tuscany. In 1592 he moved to the chair of

---

* I feel this is one of the " might have beens " of history.

Mathematics at Padua. If he had been far-seeing enough and had stayed here he would have been free of the Inquisition since Padua was under the Venetian Senate. However, when in 1613 Cosimo II of Tuscany offered him the post of First and Extraordinary Mathematician of the University of Pisa and Mathematician to his Serenest Highness, with a large salary and no duties, Galileo left Padua for Firenze. Whether he actually resigned from these high-sounding posts or was dismissed is not entirely clear. It is known that from 1613 onwards he lived either in Firenze or at his villa at Arcetri, a few miles away. He was examined by the Inquisition in 1632 and died at Arcetri in 1642 (the year Newton was born), after a career as full of achievement as any that has been known. His manuscripts and papers were all burnt by his daughter after his death.

It was at Arcetri that Galileo was visited by Milton in 1638, the year in which he, Galileo, went completely blind. At that time Galileo seems to have been under some kind of house arrest, with Cardinal Francis Barberini responsible for his good conduct, but Milton does not appear to have had any difficulty in making the visit. The aged Galileo talked to the young Milton about his astronomical discoveries, the echo of which talk is found in *Paradise Lost*. Perhaps the old man also talked of his schooldays, since Milton rode twenty miles on horseback to visit Vallombrosa, the beauty of which remained with him so that, long after, he was to write :

> Thick as autumnal leaves that strow the brooks
> In Vallombrosa, where th'Etruscan shades
> High over-arch'd imbower. . . .

Galileo's fragment on dice, *Sopra le Scoperte dei Dadi*,* was possibly written between 1613 and 1623, i.e. soon after the time when he arrived in Firenze. For Galileo, who at this time of his life and afterwards was almost entirely interested in astronomy, was busy with his telescopes while he was at Padua. After he had been a

---

* This was Galileo's own title. It is not until the collected works of 1718 that the title *Considerazione sopra il Giuoco dei Dadi* appears.

few years at Firenze we find him (1623) at the beginning of his controversies about the motion of the earth, controversies which were to occupy him for the rest of his life. By themselves these dates are not conclusive. We must add however the psychological factor. Galileo is described as being thickset, taller than the average, with a face like a pig, red-haired and possessing the fiery temper which is traditionally associated with it. The following phrase occurs in the *Thoughts about Dice-Games* :

> Now I, in order to oblige him who has ordered me to produce whatever occurs to me about the problem, will expound my ideas. . . .

It is a little difficult to imagine anyone ordering Galileo to do anything (and being obeyed) except the person who paid him, i.e. the Grand Duke of Tuscany.

Since there are no other fragments or discussions on dice-playing—as far as I know—by other mathematicians of the time, the date of this fragment, which I would guess to be between 1613 and 1623, is a matter of small importance. Galileo's style of writing is noteworthy for its clarity and also for its lack of brevity, being in fact prolix and tedious in the extreme. If there had been any doubt whatsoever about the general method of procedure in calculating chances with a die he would have given us a long exposition of it : but he does not, he plunges straight into the argument. The problem is that which has in essence been solved by Cardano. Three dice are thrown. Although there are the same number of 3-partitions of 9 as there are of 10,* yet the probability of throwing 9 in practice is less than that of throwing 10. Why is this? Galileo's remarks may be translated (see Appendix 2) as follows :

> The fact that in a dice-game certain numbers are more advantageous than others has a very obvious reason, i.e. that some are more easily and more frequently made than others, which depends on their being able to be made up with more variety of numbers. Thus a 3 and an 18, which are throws which can only be made in one way with 3 numbers (that is, the latter with 6, 6, 6, and the

---

* (621) (531) (522) (441) (432) (333) and (631) (622) (541) (532) (442) (433)

former with 1, 1, 1, and in no other way), are more difficult to
make than for example, 6 or 7, which can be made in several
ways, that is a 6 with 1, 2, 3, and with 2, 2, 2 and with 1, 1, 4 and
a 7 with 1, 1, 5 ; 1, 2, 4 ; 1, 3, 3 ; 2, 2, 3. Nevertheless, although 9
and 12 can be made up in as many ways as 10 and 11, and therefore
they should be considered as being of equal utility to these,
yet it is known that long observation has made dice-players
consider 10 and 11 to be more advantageous than 9 and 12.

This extract serves to show how he begins the topic assuming
that partitional calculations were known, as indeed they had been,
at least since the commentary on *De Vetula*. (It also serves to
illustrate the prolixity of Galileo's style, a prolixity which he
shares with many of the early mathematicians and which may
well have arisen from the mental struggle which they had to put
their ideas on paper.) After some discussion of the six 3-partitions
of 9 and of 10, he goes on :

> . . . since a die has six faces, and when thrown it can equally well
> fall on any one of these, only six throws can be made with it, each
> different from all the others. But if together with the first die
> we throw a second, which also has six faces, we can make 36 throws
> each different from all the others, since each face of the first die
> can be combined with each of the second. . . .

He says that the total number of possible throws with three dice
is 216 and gives a table of possible throws for totals of 10, 9, 8, 7,
6, 5, 4, 3, noting that the totals for the numbers 11 – 18 are
symmetrical with these. Thus from his table the number of
possible throws for 10 is 27 and for 9, 25. His treatment of the
problem is exactly that which we should use in an elementary
attack today and leaves no doubt that the calculation of a
probability from the mathematical concept of equally-likely sides
of the die was clearly known to at least one Italian mathematician
of the early seventeenth century. The person (His Serenest
Highness?) asking Galileo the question had gambled often enough
to be able to detect a difference in probabilities of 1/108.

That Galileo had to a certain extent to occupy himself with
trivialities while he was mathematician to his Serenest Highness

is illustrated by the letters which he exchanged in 1627 with a priest called Nozzolini. These letters were printed in 1718 in the collected works of Galileo, and the Editor of these letters gives the following amusing preamble :

> It was the custom in those happy times in our city of Firenze to hold various literary gatherings in the houses of those great lords who did not spend their time either running after women or in the stables or in excessive gambling, but rather passed their day in learned discussions and among educated people. In one of these discussions the following problem was posed : A horse which was really worth 100 scudi was by one man valued at 1000 and by another at 10 : which of the two was the better estimate and which of the two judged more extravagantly? There was living then a certain Nozzolini, a priest, and a cultivated man Andrea Gerini, a Florentine gentleman, sent the problem to this Nozzolini : the latter gave his judgement on the question which was that to make the estimate one should use arithmetic and not geometric progression and that he considered the man who had valued the horse at 1000 scudi had been more extravagant in his estimate than he who had valued it only at 10.

Galileo was asked about this and he started a disputation with Nozzolini maintaining that a geometric progression should be used. The correspondence is voluminous but is noteworthy, perhaps, only as an illustration of the clarity of Galileo's mind as is witnessed by the way in which he starts the argument. His first letter begins :

> To arrive at a conclusion about the subject under consideration, which is, which of two valuers has valued the better and less extravagantly a thing which is really worth 100, the one who values it at 1000 or the one who values it at 10, it seems to me that one must first establish what is considered a true and just estimate and what is considered an unjust and extravagant one. The man who values at 100 a thing which is really worth 100 will value justly and well ; those who value it at less or more will deviate from the proper valuation. And among these he who deviates most from the proper valuation either above it or below it, will be guilty of the greatest extravagance. And since some will perhaps consider that to

deviate equally from the truth, both above and below, may be understood in two ways, that is either by arithmetical proportion—which is when the excess of the estimate over the proper one is equal to the excess of the proper estimate over the lowest valuation, e.g. if the proper estimate was 10 and one calculation was 12 and the other 8 then both guesses are equally good—or by geometrical proportion which is when the higher estimate has the same proportion to the real one as the real one has to the lower estimate . . . .

Galileo had been well trained by the Jesuits in hair-splitting logic* but does not contribute anything new as far as estimation is concerned, except for the recognition that opinion can, or should be, weighed in objective fashion.

Galileo was born twelve years before Cardano died. The difference between the writings of these two men is illustrative of the speed with which human thought was developing at that time. Cardano was, in thought, still linked with the Dark Ages with its suffocating atmosphere of superstition. Galileo is of our time, for although his style of writing is irritating in its prolixity and tortuousness of argument he obviously thinks in a way not too different from our own. As a contributor to the calculus of probability he is negligible, but as an indicator of the state of that calculus and of scientific thought generally he is of great importance.

He died in January 1642, and although he asked to be buried in his family vault in the Church of Santa Croce in Firenze the Pope and the Inquisitors did not permit it. He was buried privately in a corner of that church. Here also lie Michelangelo and Macchiavelli.

---

* Perhaps the effect of the foundation of the Order of Jesus in 1534 on scientific method in the sixteenth century should be a subject for study. One is reminded of Charles Kingsley's gibe, " Truth, for its own sake, had never been a virtue with the Roman clergy."

References

Hoeffer's *Nouvelle Biographie Universelle* devotes some pages to Galileo's life and work.

Montucla, *Histoire des Mathématiques*

and

Libri, *Histoire des Sciences mathématiques en Italie*

are useful references here. A modern appreciation by

Sherwood Taylor, *Galileo and Freedom of Thought*

is excellent in that the intellectual spirit of the man emerges. I have made much use of the vast definitive edition (20 volumes) of

Galileo, *Opera Omnia*.

There are many biographies of Galileo, but they treat him as an astronomer and natural scientist, as is obvious that they would.

chapter 8

Fermat and Pascal

> For a brief space it is granted to us, if we will, to enlighten
> the darkness that surrounds our path . . . . We press forward,
> torch in hand, along the path. Soon from behind comes the
> runner who will outpace us. All our skill lies in giving into
> his hand the living torch, bright and unflickering, as we
> ourselves disappear in the darkness.
>
> HAVELOCK ELLIS

The flood of ideas generated by the Italian Renaissance did not
abate after the death of Galileo, but it was apparent even in
Galileo's lifetime that pre-eminence in both the arts and sciences
was passing from Italy to France. The Renaissance in France was
later than that of Italy and was probably greatly helped by the
loot in the form of ideas, manuscripts and books which successive
French armies carried out of Italy. The fabulous library of
manuscripts of Petrarch is said to have been looted by the French
in 1501, and this must have been only one of many instances. By
the year 1600 the revolution in the arts and the development of
the sciences had both reached mature proportions, although they
did not produce any practitioner of the experimental method of
the calibre of Galileo or Leonardo. The major contributions of
the French scientists of the seventeenth century, when one surveys
all the work done, might be said to be in the realms of pure and of
applicable mathematics.

Galileo wrote as though the calculation of a probability was
something which was obvious, and the suggestion from his work
is that almost any mathematician could set out the method.
Although this fragment was left unpublished by him, it does
appear likely that the calculation of a probability *was* a common-

place to the Italian mathematicians and probably therefore to some of those in France. Thus, while the foundation of the calculus of probabilities, the giant step forward made with the concept of the equally-likely faces of the die, does not belong to the French, the next development of the theory is undoubtedly theirs. And by far the greatest of the French mathematicians of the seventeenth century was Pierre de Fermat.

Fermat, born in 1601 at Beaumont-de-Lomagne near Montauban in Gascony—five years after Descartes and four years before Rembrandt—has been called by some the prince of amateurs and by others the greatest pure mathematician who has ever lived. There can be no doubt that, in an age when mathematical theories in general were being developed at a great rate, he was outstanding, and this quiet lawyer did more than any other Frenchman in helping the formulation of the theory of probability. Fermat was the son of Dominique Fermat, citizen and second consul of the town of Beaumont, and his wife Françoise de Cazelneuve. His life offers few noticeable incidents. His family were leather merchants and he spent his childhood at home. He studied law at the University of Toulouse and after passing his examinations was named *Conseilleur de la Chambre des Requêtes du Parlement* of Toulouse in May, 1631, at the age of 30. Some days after he was given this position he married Louise du Long, daughter of a Counsellor in the same Parlement, and it is from this date that he assumed the prefix *de*. It is not known whether he was actually ennobled by a special decree or whether the post of Counsellor carried the implicit right to the prefix.

In the intervals when the law courts went into recess he studied both literature and mathematics. Parliamentary Counsellors of that age, like the judges of today, were obliged to hold themselves aloof from and to have very little contact with, their fellow citizens, and this must have helped to produce the necessary time for reflection. His biographers speak of his " singular erudition " in what would now be called the humanities. His knowledge of the chief European languages and of the literature of continental Europe was said to be both wide and accurate. He made emendations to Greek and Latin texts. He wrote verses in

Latin, French and Spanish and he did research in mathematics. This quiet friendly man passed all his working life in the service of the State. He was promoted King's Counsellor, still in the Parlement of Toulouse, in 1648, and he died at Castres in January 1665, when he was 64 years of age. His private life seems to have been as uneventful and successful as his public career. He had two daughters, both of whom became nuns, and three sons one of whom, Samuel, became known as a writer. Even after three hundred years the good temper, modesty, kindliness and intellectual brilliance of this great man shine through his letters which have now, fortunately, been collated, dated and printed in full.

Unlike the Italians of the sixteenth and seventeenth centuries and the French, Swiss and Italians of the eighteenth century, the public challenge to the problem does not seem to have played much part among Fermat and his contemporaries. It has been mentioned how Fiore, the pupil of Ferreo, gained a great reputation for himself and his teacher by challenging other mathematicians to solve problems concerning the roots of a cubic equation, and later John Bernoulli used this device to further his own reputation, but public disputation does not seem to enter just at this time. It would appear rather that the solution of a problem was followed by a letter to a friend telling him about it, and, possibly just to puzzle him a little, a step in the proof is held back, but this was done privately. Letters were also exchanged setting out the failure to solve a problem and asking for enlightenment. In all the vast correspondence of Fermat there appears only one suggestion that the correspondent might have been resentful of Fermat succeeding where he himself had failed. This correspondent was Pascal. The development of mathematics in seventeenth-century France is interesting in that so very little was made public in comparison with what was achieved. If a scientist belonged to the closed circle then he corresponded with those others of the circle about anything and everything : but the comments and approbation of his equals seem to have been sufficient for him, and it was unusual for their letters to find their way into print. It is only because so many of these letters have survived that it is possible nowadays to give Fermat the credit which is certainly his right. On Fermat's part the lack of desire to publish may have been his modesty. He did not have the day-to-day contact which the group of mathematicians had in Paris and which must have supplemented their letters to each other. He seems to have been one of those rare persons—like Newton—who flourished in isolation and who was modest enough to believe that the sketch of a proof or even the statement of a theorem was enough for all the world to understand.

Fermat carried on an immense correspondence with scientists in Paris, but there are also records of letters going to the Low Countries, to Italy and sometimes to England. This century, the seventeenth, is noteworthy for the formal founding of a great number of scientific societies, and in this correspondence between the scientists of all nations we have the nuclei of these. Harcourt-Brown states that the Italian academies or societies date from the fifteenth century, and possibly the French borrowed this idea from them ; or the societies may just have been the inevitable consequence of the liberation of thought following the slow climb of humanity from the Dark Ages. In France the academies were not so much social (as in Italy) as the regular meetings of scientists with comparable interests who exchanged ideas about the scientific problems of the day. At about the time of Fermat's birth one of the links between the academies of Italy and Paris was Nicolas-Claude Fabri de Peiresc (1580–1637). He is described as being inquisitive and very curious and was the ideal person to spread the gossip of new ideas and techniques. He had been a student at Padua and had contacts with the scholars of Firenze and other Italian centres of learning. During the years 1605–1620 he travelled extensively all over Europe, spreading his news as he went, and since one of his loves was astronomy he would have given news of Galileo. He also carried on a lively correspondence with the Abbé Mersenne. From the point of view of the probability calculus he is interesting only as a typical example of the way in which mathematical and other ideas were carried about Europe and because of his contact with the Mersenne Academy. Yet another possibility whereby the ideas about chance may have

been propagated is Cardinal Francis Barberini who was liberal
in his interpretation of Galileo's house-arrest and who was also
a correspondent of the Abbé. So many of the threads from divers
points in Italy are tied together on their arrival at the Academy
of the Abbé Mersenne that the probabilities would be in favour
of ideas about chance having come to this group from Italy unless
they had arrived earlier and before its foundation. The group
is a noteworthy one.

Marin Mersenne* was born in 1588 at the village of La
Soultière, near Oizé (Sarthe), and educated at the Jesuit Seminary
at La Flèche.† He became a priest in the order of the Minorites
and was stationed in Paris for much of his life. It is said that
from his earliest years he was interested in music, in mathematics
and in the natural sciences and he corresponded with nearly every
scientist of note of his day. Harcourt-Brown writes :

> There is hardly a figure of importance in the learned world who
> does not appear in the pages of his letters. From all parts of
> Europe news of the advancement of the sciences came to the
> convent " des pères Minimes, proche la Place Royale " and thence
> went the prized letters of the reverend father, written in their own
> peculiar, cramped and all but illegible hand, with the precious
> news of Descartes, Morin, Fermat, Torricelli or Galileo.

To Mersenne's house once a week came mathematicians and
natural scientists, among them Gassendi, Desargues, Carcavi,
Roberval, Descartes and the Pascals, father and son. The bond
appears to have been Mersenne.

> All who came into contact with him remarked on his universal
> learning, on the sweetness and charm of his speech, the gentleness
> of his temper, the *naïveté* which won its way into all hearts.

Mersenne died in 1648 and the group changed its venue to the
house of La Pailleur. After he died in 1651 they met in other

---

* A translation of a biography of Mersenne is given as Appendix 3. It is
particularly interesting to note the extensive list of his correspondents.

† Descartes (1596-1650), born at La Haye in Touraine, was educated at the
same seminary. It would be interesting to know who was the Jesuit teacher
of mathematics and natural sciences who produced two such able pupils.

places. These informal societies died with the formal beginnings of the learned societies, the Royal Society of London in 1660 and the Académie des Sciences of Paris in 1665. It would seem likely that Fermat was *au courant* with all that was being discussed in the scientific world both through his own correspondence and through the information which he would gain through the Mersenne Academy. It is almost certain that if the fragment of Galileo on dicing had been thought sufficiently new to discuss, the informal academies would have done so within a few years of its having been written.

The name of Blaise Pascal is always linked with that of Fermat as one of the " joint discoverers " of the probability calculus. Because his mathematical work came in bursts before he retired at an early age to meditate on " the greatness and the misery of man " and was negatively correlated with these meditations, it is worthwhile to consider briefly the outline of his life. Étienne Pascal was a judge at Clermont-Ferrand in the Auvergne and is described as a very learned man and an able mathematician. In 1631, having, so his biographers say, " an extraordinary tenderness for his child, his only son," he gave up his post as judge and moved to Paris in order to supervise his son's education. He became a member of Mersenne's Academy, although little is recorded of his part in their discussions. Many sources agree, however, that he did take an active part. Carcavi, who, like Fermat, was a Counsellor in the Parliament of Toulouse, was a correspondent of Mersenne before he moved to Paris in 1642 and became a member of the Academy. He was responsible for introducing Fermat into the correspondence circle of Mersenne in 1636, and it was at his suggestion that Étienne Pascal and Roberval wrote to Fermat (April 26th, 1636) concerning the weight of the earth. Roberval and Pascal attacked Fermat's theory and there was an exchange of letters about it. The relations of Fermat with the Academy remained, however, excellent, and when in 1637 Descartes attacked Fermat's method of maximum and minimum tangents Roberval and Pascal were Fermat's defenders, supporting him with a polemic which is well known. The elder Pascal introduced Blaise into the academy when he was fourteen years old, that is, about the year 1637 when the controversy was at its height.

Blaise Pascal (1623–1662), born when Descartes was 27 years of age, appears to have been of poor physique and of precocious mental ability. His father, it is stated, from an early age concentrated on developing his reasoning powers rather than his memory. Blaise was undoubtedly quick-witted. His sister, married to a physicist, M. Périer, wrote a life of her brother which is not without a certain imaginative interest. She says her brother had an extraordinary wit at a very young age which he showed by repartee quick and to the point. She also relates that Étienne Pascal was afraid of his son overtaxing his strength, and while he instructed him in what is described as the usual education of the time (presumably classical languages, commentaries on Plato and Aristotle and so on), he hid all books on mathematics from his son. One day he found him drawing diagrams in charcoal and saw that he had rediscovered for himself Euclid's propositions.* Now although Madame Gilberte Périer was imaginative with regard to her brother and while much that she writes may be discounted for this reason, there is no doubt that Blaise could invent for himself without prompting of book or person. For when he was sixteen (c. 1639) he wrote his famous *Essai pour les Coniques* which, although no longer surviving as a whole, certainly did exist since Leibnitz reports having seen it. Blaise's work caused a certain amount of controversy ; some mathematicians received it with acclamation, while others, among them Descartes, refused to believe that it was entirely his own work. While granting more to Pascal than Descartes was willing to do, it may be that he was right to be a little sceptical. The method of projection used had been put forward previously by Desargues, and while the creative thought for the essay probably came from Blaise we may wonder how far Étienne acted as an improver or refiner of his son's work.†

---

* It is the hallmark of a mathematical prodigy to rediscover Euclid's propositions and *de rigueur* to recount this in his biography, certainly in the seventeenth and eighteenth centuries and probably earlier. Euclid lived B.C. 306-283 and one has the suspicion that the story originated with the first mathematical prodigy of the next generation to him.

† The parallel of the boy Mozart playing his melodies and his father writing them down and " improving " them is irresistible.

The maturity of the essay, which caused Descartes to be suspicious of its authorship, was possibly due to Étienne. But even so, allowing for his father's improvements and Desargue's trail-breaking in the form of method, enough is left to demonstrate the astonishing mathematical powers of this boy of sixteen.

Two years later Blaise invented a calculating machine, but his health, already delicate, deteriorated and he had to give up working for four years. This is possibly the worst thing which could have happened to this delicate introspective boy who wanted to know the reason of everything, and who, when adequate reasons were not given him, looked for better ones for himself. Four or five years after his breakdown, in 1646, we hear of his conversion to Jansenism in company with the rest of his family. This cult of Jansenism affected many scientists of the seventeenth century and was the ostensible* reason for the failure of Pascal to fulfil his mathematical promise, so I will consider the history of the cult briefly and its destiny. Without attempting to dis-entangle the theological doctrines involved, and thereby being quite unjust, it would appear that the cult of Jansenism arose from a violent dislike (and possibly envy) of various Catholic bishops and heads of seminaries for the power of the Society of Jesus. This Society, founded in 1534, was growing all-powerful, and many Catholics of importance felt dislike of the rate at which its in-fluence was increasing. Within the framework of Catholic dogma, therefore, they set themselves to attack the Jesuits on the subjects of freewill and of the grace of God. The Jansenists took what has been described as a standpoint akin to Calvinism. Cornelius Jansen, created Bishop of Ypres in 1636, was the leader of this sect and it was the tract written by Jansen on *Réformation de l'Homme Intérieur* which is held to have been the cause of Pascal's conversion. Blaise appears to have become interested in the sect when his father was nursed by the Jansenists during an illness at Rouen, and he finally retired to die at Port-Royal. The celebrated Cistercian Abbey of Port-Royal, built in the valley of the Yvette

---

* I write " ostensible " because Pascal was so tied by the loving care of his father and his sisters and by his own bodily weakness that some interruption of this kind seems inevitable.

30 miles west of Paris in the village of Les Hameaux, was founded in 1204 by the wife of a French nobleman when he was absent from France on the Fourth Crusade. The Pope gave this abbey the privilege of affording a retreat to lay persons who wanted to withdraw from the world for a time but who did not want to bind themselves with permanent vows. A second abbey of the same foundation was instituted in Paris in 1626, and it was to this Abbey that Pascal retired for meditation, the last time for good.

Jansen died in 1638 and his final apologia was printed by his friends after his death (1640). The real struggle between the Jesuits and the Jansenists now began, with the Jesuits trying to persuade successive popes to declare the Jansenists heretics and to excommunicate them, and with the Jansenists preaching and pamphleteering against the Jesuits. The Sorbonne as a whole seems to have been moderately inclined to Jansenism. Bishop Jansen in his *Discours* held that scientific curiosity was only another form of sexual indulgence. " On reading this page," writes Sainte-Beuve, an eminent biographer of Pascal, " a curtain was drawn in Pascal's soul. Physics, geometry appeared to him for the first time in a new light." This first conversion of Blaise in 1646, when he was 23, does not seem to have lasted very long, for in 1648 he took up mathematics and physics again. Périer, at Pascal's suggestion, carried out the famous experiments with a baro- meter at the Puy de Dôme. The conclusions which Blaise drew from these experiments he wrote in *La Pesanteur de l'Air* and thus involved himself in further dispute with Descartes. Descartes had discussed the possibility of these experiments in letters to Mersenne, and Pascal would have been privy to these since he was a more or less regular attendant at the academy. The scepticism of Descartes with reference to Blaise's essay on conics has already been noticed. He was put out by these experiments of Périer's, and this was probably accentuated by the fact that Descartes (like Mersenne) was a protégé of the Jesuits and a great lover of the Society of Jesus. In the long run he does not seem to have borne Blaise any malice, but the incident seems to show that Blaise is not entirely the originator which posterity is inclined to believe him. After the controversy Blaise wrote to his sister Jacqueline that as a result of his terrible struggle and the indulgence of his scientific curiosity he had become paralysed and was able to walk only with crutches. This paralysis did not however last long.

In 1651 Étienne died, leaving Blaise a moderate fortune : one of his sisters was married and the other had become a nun. The leading strings were at last removed and in 1653 he is described as a man of the world, leading a dissolute life. Whether he was exceptionally wild or whether he merely led the life of any young man of that time it is not possible to say, but whatever the truth of the matter his delicate health (probably) brought him up short before many months had elapsed. In 1654 he had the famous correspondence with Fermat on the problem of Points. Letters passed between the two scientists during the four months July to October, and the correspondence was definitely closed by Pascal before his second conversion on November 23rd, 1654. It is said that the horses of a four-in-hand ran away with him : he took this as a sign from God that he must give up the life he was leading and do no more mathematics, and he retired to Port-Royal. The story of the rest of his life is concerned more with Jansenism than with mathematics. At Port-Royal he could not still his restless enquiring mind and before long he took up the cudgels for the Jansenists against Pope Innocent X. The Sorbonne had now decided that it was politic to regard the Jansenists as heretics. Pascal published in January 1656 the first two of his famous Provincial letters, of which Voltaire wrote : " A book of genius is seen in *Les Lettres Provinciales*. All types of eloquence are to be found in them : there is not a single word which after 100 years should be changed." Voltaire was, however, anti-Jesuit himself. At the time (1656) the letters made a great impression but could not save the Professor of Theology (Antoine d'Arnauld) from being expelled from the Sorbonne.

The Jansenists were persecuted, their leaders were forced to go into hiding, and the nuns of Port-Royal were subjected to imprisonment. Pascal continued to live in extremely ascetic circumstances, spending his time reading the Scriptures and in writing down the thoughts which these spiritual exercises evoked. These thoughts were published after his death and form the famous *Pensées*. It is related that in 1658 he had toothache which kept him awake, and that to distract himself he thought about the cycloid, the curve traced out by a fixed point on the circumference of a wheel rolling at a uniform speed on a horizontal plane. As he thought, the pain disappeared. He took this to be a sign that the Almighty didn't mind him thinking about the cycloid, so he thought about it for eight days and wrote his results to Carcavi. The fact that he wrote under a pseudonym is possibly a sign that he did not want his world to know that he had fallen so far from grace as to indulge in the sin of mathematical research. He died in 1662, at the age of 39, from convulsions, the post-mortem showing that he had a serious lesion in the brain. He was buried in the church of St. Etienne-du-Mort.

References

In this and succeeding chapters I have gained much information from

> HARCOURT-BROWN, *Scientific Organisations in France and Italy in the 17th Century*.

There is a great deal in this book which is new and not easily accessible elsewhere. For the lives of Pascal and Fermat I have used the appropriate volumes of Hoeffer, *Biographie Universelle*, checking and supplementing the information both from the references given and, in the case of Pascal, from the commentaries written to accompany the reprinting of his complete works,

> B. PASCAL, *Oeuvres* (v. 14) *pub. suivant l'ordre chronologique, avec documents complémentaires, introductions et notes*.

These commentaries are fully documented and form a most valuable contribution to one's knowledge of Pascal. Harcourt-Brown devotes a little space to Mersenne. One may supplement this by referring to his correspondence which is now in the course of being printed. Fermat's letters were reprinted in 1894 and the editors contribute exhaustive footnotes.

chapter 9

The arithmetic triangle and correspondence between Fermat and Pascal

L'homme n'est qu'un roseau, le plus faible de la nature,
mais c'est un roseau pensant.

PASCAL, *Pensées*, vi, 347.

It has been previously noted that the arithmetic triangle commonly attributed to Pascal is of much earlier provenance. Michael Stifel gave the table

```
1
2 - 1
3 - 3
4 - 6
5 - 10 - 10
6 - 15 - 20
7 - 21 - 35 - 35.
```

Stifel was interested in the triangle in order to find a practical method to extract roots of various orders. This was also the main purpose of Tartaglia (*Generale Trattato*, 1556) and Simon Steven of Bruges (*L'Arithmétique*, 1625). Mersenne was interested in the theory of combinations and discusses the theory in at least three places in his books, *La Verité des Sciences* (1625), Book III, Chapter 10, *Harmonicorum*, and *Harmonie Universelle*. He cites the calculations of Xenocrates but otherwise gives no reference except to an unknown individual of whom he speaks mysteriously and whom he designates by the letters I.M.D.M.I.* From some fragmentary correspondence left by Mersenne it would appear that Aimé de Gagnières was interested in his combinatorial calculations, as also was the mathematician Frenicle, although his work did not appear until much later (*Abrégé des Combinations*, 1693). That

---

* It has been suggested that M.D.M. stands for Monsieur de Méré but this is unlikely.

Pascal was not original is, however, quite definite when it is remembered that Hérigone is supposed to have been his teacher. Hérigone in his *Cours mathématique* (Paris, 1634) constructed a table of numbers with the idea of calculating the coefficients of integer binomial powers.   He proposed—

$$
\begin{array}{ccccccccc}
 & & & & 1 & & & & a \\
 & & & 1 & 2 & 1 & & & a^2 \\
 & & 1 & 3 & 3 & 1 & & & a^3 \\
 & 1 & 4 & 6 & 4 & 1 & & & a^4 \\
1 & 5 & 10 & 10 & 5 & 1 & & & a^5
\end{array}
$$

and he devoted a chapter to combinatorial calculations (*Arithmétique pratique*, Chapter XV, " Des diverses conjonctions et transpositions," Tome II, p. 119).   Pascal cites the works of Hérigone at the end of his own treatise *Usage du Triangle Arithmétique pour trouver les puissances des binômes et apotômes*.   He possibly knew of the work of Mersenne; he certainly knew of the work of Gagnières since he refers to it. This all suggests that it would be more appropriate to speak of the " precious mirror of the four elements " rather than " Pascal's arithmetic triangle " for he was very nearly the last of a long line of "discoverers".

He mentioned the arithmetic triangle in a letter to Fermat in August 1654.   Pierre de Carcavi first put Fermat and Pascal in touch with one another.   It will be remembered that he played the role of intermediary between Etienne Pascal and Fermat in 1636.   Throughout his life he seems to have played a useful part in introducing scientists from outside France and from the French provinces to those scientists whom he met at the Mersenne Academy.   Carcavi had known Fermat when he was still at Toulouse, and he was an intimate friend of Blaise Pascal.   In *La Vie de M. Descartes* (1649) we come across the statement " M. Pascal had no friend more intimate than Carcavi, not excepting even M. de Roberval or the Gentlemen of Port-Royal."

Not all the correspondence between Pascal and Fermat has survived.* The first letter from Pascal to Fermat is missing. The

---

* A complete translation of the text of such letters as we have is in the Appendix. When this translation had been completed our attention was drawn to D. E. Smith, *A Source Book of Mathematics*, in which a translation of the letters is given by V. Sandford.

order of some of the others has been altered, but because of the untiring efforts of the editors of Fermat's papers, we have what is obviously the reply to this first lost letter. It was originally placed in the middle of the series but clearly belongs at the beginning, having regard to its content. All the letters are about the problem of points. Pascal's first letter was almost certainly concerned with a gambler undertaking to throw a six with a die in eight throws. Suppose he had made three throws without success, what proportion of the stake should he have on condition he gives up his fourth throw? Fermat replies :

Sir, if I undertake to make a point with a single die in 8 throws and if we agree, after the stakes are made, that I shall not play the first throw, then I should take from the stakes one-sixth of the total as recompense for giving up the first throw. And if we agree after this that I shall not make the second throw, I should, for my indemnity, draw one-sixth of the remainder which is 5/36ths of the total. And if we agree after this that I shall not play a third time I should draw one-sixth of the remainder which is 25/216ths of the total. And if after this we agree I shall not play a fourth time, I must draw one-sixth of the remainder which is 125/1296ths of the total and I agree with you that this is the value of the fourth throw, supposing one has already settled for the preceding ones. But you propose to me in the last example of your letter (I quote your own words) : " If I undertake to throw a 6 in 8 throws and I have played 3 without success, if my opponent proposes to me that I should not play my fourth throw and he wants to pay me off because I want to make it, there will come to me 125/1296ths of the total of our stakes."

This is, however, not true according to my principle. For in this case, the first three throws having brought nothing to the die-caster, the whole of the stake remains in the game, and he who throws the die and who agrees not to play the fourth throw must take for recompense one-sixth of the total. And if he had played four throws without achieving the looked-for point and we agree that he should not play the fifth, he will still have the same one-sixth of the total for his indemnity. For the entire sum remaining in the game, it does not follow only from principle but it is common sense that each throw should give equal advantage. I beg you therefore to tell me if we agree in principle, as I believe, and if we differ only in application.

Pascal replied to this on July 29th, 1654. From the tone of his letter Fermat's answer to the first letter cannot have been written much before this. Succeeding letters between the two have about two weeks between them. Pascal's letter begins :

Sir, I am impatient as well as you and although I am still in bed I cannot prevent myself from telling you that I received yesterday evening from M. de Carcavi your letter about the die game which I admire more than I am able to tell you. I have not the leisure to say much but in a word what you have done is absolutely right. I am quite satisfied with it, for I do not doubt any more that I have the truth after the admirable agreement I find with you . . . . I have seen several people obtain that for dice, like M. le Chevalier de Méré, who first posed these problems to me and also M. de Roberval. But M. de Méré has never found the true value for the division of the stakes nor the method of deriving it, so that I find myself alone in discovering this.

Your method is very sure and is that which came to me the first time I thought about this problem : but because the labour of combinations is excessive, I have found a shorter way which I will tell you briefly. For I would like to open my mind to you, so much pleasure has our agreement given me. I see that truth is the same in Toulouse and in Paris.

It is perhaps not unfair to comment here that if the editors of Fermat's letters are correct in this ordering of the sequence of letters, and it is generally acknowledged that they are, then Pascal may be a trifle hypocritical. Possibly Fermat had mis-understood what Pascal meant in the first (missing) letter, but it would seem that Blaise had not got the right solution to the prob-lem which he posed and that he possibly only arrived at it after the receipt of Fermat's letter. The fact that in this answering letter of July 29th he goes on to give the solution to another problem does not signify much since Pascal was extremely quick at algebra when headed in the right direction.

The Chevalier de Méré has become famous as the gambler whose questions started the Pascal–Fermat correspondence, but little is written about him. He seems to have carried the Problem of Points to many of the Paris mathematicians including Roberval. In a primitive sort of way, and to a limited extent, he appears to have dabbled in mathematics himself, although his collected papers *Les Oeuvres de Monsieur le Chevalier de Méré* (Amsterdam, 1692) bear no evidence of this, being literary pieces of not very high calibre. That he was not without the good opinion of himself that often goes with a second-rate intelligence can be seen in a letter written by him to Pascal some time after 1656.

> " You must realise," he writes, " that I have discovered in mathematics things so rare that the most learned of ancient times have never thought of them and by which the best mathematicians in Europe have been surprised. You have written on my inventions as well as M. de Huyghens, M. de Fermat and others."

All he can mean by this is that he asked the original question. It is true that Leibnitz wrote of him : " The Chevalier had an extraordinary genius for mathematics," but since this occurs in a passage where Leibnitz was being derogatory about him he may only have intended sarcasm. It has never been stated how Pascal and de Méré began to discuss such problems, but it may be noted that this contact was made during what might be called Pascal's dissolute period. Since de Méré was not a member of the continuance of what had been the Mersenne Academy, they may have met in less academic surroundings.

Pascal's letter to Fermat of July 29th goes on to discuss the following problem. Two players play a game of three points and each player has staked 32 pistoles. How should the sum be divided if they break off at any stage? (Throughout it is assumed by both Pascal and Fermat that the players are of equal skill and opportunity.) Pascal writes :

> Suppose that the first player has gained 2 points and the second player 1 point. They now have to play for a point on this condition, that if the first player wins he takes all the money which is at stake, namely 64 pistoles, and if the second player wins each player has 2 points, so they are on terms of equality, and if they leave off playing each ought to take 32 pistoles.* Thus, if the first

---

* The argument is reminiscent of that of Peverone.

player wins, 64 pistoles belong to him, and if he loses, 32 pistoles belong to him. If then the players do not wish to play this game, but to separate without playing it, the first player would say to the second : " I am certain of 32 pistoles even if I lose this game, and as for the other 32 pistoles perhaps I shall have them and perhaps you will have them ; the chances are equal. Let us divide these 32 pistoles equally and give me also the 32 pistoles of which I am certain." Thus the first player will have 48 pistoles and the second 16.

He now supposes that the first player has gained two points and the second player none, and that they are about to play for a point. The condition is then that if the first player gains this point he wins the game and takes 64 pistoles, and if the second player gains the point they are in the position already examined, in which the first player is entitled to 48 and the second to 16. Thus, if they do not wish to play, the first player could say to the second : " If I gain the point I win 64 and if I lose it I am entitled to 48. Give me the 48 of which I am certain and divide the other 16 equally since our chances of gaining this point are equal." Thus the first player will have 56 and the second 8 pistoles. Finally suppose the first player has gained one point and the second none. If they proceed to play for a further point and the first player wins they will be in the condition already examined (56 : 8) while if the second player wins they will have a point each and be entitled to divide (32 : 32). Thus if they do not play this point the first player could say : " Give me 32 and divide 56–32 equally," so that he would be entitled to $32 + 12 = 44$. This is the substance of what Pascal writes to Fermat on the specific problem they are discussing, and so far there is nothing new. He now goes on to generalise it, possibly a trifle insecurely.

> Now to make no mystery of it, since you see it so well (I put out everything clearly just to see that I had made no mistake), the value—by which I mean only the value of the opponent's money —of the last game of *two* is double that of the *last* game of three and *four* times that of the last game of four and eight times the last game of *five*, etc. But the proportion for the first games is not so easy to find ; it is as follows, for I do not wish to conceal anything. Here is

the problem which I made so much of because it pleases me greatly. It is :—Being given as many games as you wish, find the value of the first. Let the number of given matches be, for example, 8. Take the first eight even numbers and the first eight odd numbers, thus

$$2, 4, 6, 8, 10, 12, 14, 16$$
$$1, 3, 5, 7, \quad 9, 11, 13, 15.$$

Multiply the even numbers in the following way : the first by the second, the product by the third, the product by the fourth, the product by the fifth, etc. Multiply the odd numbers in the same way, the first by the second, the product by the third, etc. The last product of the even numbers is the denominator and the last product of the odd numbers is the numerator of the fraction which expresses the value of the first of 8 games.* If each person stakes the number of pistoles expressed by the product of the even numbers, he would get from his opponent's stake the product of the odd numbers. This can be shown, but with a great deal of trouble, by the theory of combinations, as you have worked out, and I have not been able to demonstrate this by any other method but only by combinations. And here are the propositions which lead to it, which are really arithmetical properties bearing on combinations which I find have certain beautiful properties :—If of any eight letters taken at random, say ABCDEFGH, if you add one half the combinations of 4, i.e. 35 (half 70) with all the combinations of 5, namely 56, plus all the combinations of 6, namely 28, plus all the combinations of 7, namely 8, plus all the combinations of 8 you get the fourth number in the fourth progression whose origin is 2† . . . .

This is the first proposition, which is merely arithmetic. The other concerns the problem of points and is as follows. It is necessary to say first of all that if I win the first point out of 5 and thus need 4 more, the game must certainly be decided in 8 which is twice four. The value of the first throw of a set of 5, in terms of your opponent's stakes, is the fraction which has for

---

* i.e. n games $\dfrac{1.3.5. \ldots (2n - 1)}{2.4.6. \ldots 2n} = \dfrac{(2n - 1)!}{n!(n - 1)!} \dfrac{1}{2^{2n-1}}$

† 2, 2^3, 2^5, $(2^7 = 128)$. $35 + 56 + 28 + 8 + 1 = 128$. Generally, given $2n$ letters, owing to the symmetry of the binomial, $\frac{1}{2}(1 + 1)^{2n} = 2^{2n-1}$.

numerator one half the number of combinations of 4 on 8 (I take 4 because it is equal to the number of points required and 8 because it is double 4), and for denominator this same numerator plus all the higher combinatorial values. Thus if I have won the first game out of a set of 5, there comes to me from my opponent's stakes 35/128 . . . and this fraction 35/128 is the same as 105/384 which we get by the multiplication of even numbers for the denominator and odd numbers for the numerator. You will doubtless understand this well if you think about it which is why I shan't bother to explain it any more.

He gives various tables which indicate how the stakes should be divided.

Had Pascal left it at this, one would have marvelled at the quick intelligence which was able to generalise from such small arithmetical examples. The continuation of this present letter, however, and of succeeding letters does to a certain extent throw doubt as to whether he really understood what he was about. After these tables for the division of the stakes and a further discussion of the points problem he continues :

I haven't time to send you a solution of a difficulty which has puzzled M. de Méré. He has a good intelligence but he isn't a geometer and this, as you realise, is a bad fault. He does not understand even that a mathematical line is infinitely divisible and holds very strongly that it is composed of a finite number of points and never have I been able to dissuade him of this.*

The letter proceeds in possibly ingenuous fashion :

If you are able to solve the difficulty it would be perfect. The Chevalier de Méré said to me that he has found falsehood in the theory of numbers for the following reason. If I undertake to throw a six with one die, there is an advantage in undertaking to do it in 4 throws, as 671 to 625. If I undertake to throw the

---

* It is because of this last sentence that one is able to name the Chevalier de Méré as the proposer of the problem of points. The name was left as M. de M . . . in the letters from Pascal to Fermat, but there are letters between Pascal and the Chevalier de Méré in which this point of the infinite divisibility of a line is argued.

" Sonnez "* with two dice there is a disadvantage in undertaking to do it in 24.† And moreover 24 is to 36 (which is the number of pairings of the faces of two dice) as 4 is to 6 (which is the number of faces of one die). This is his " grande scandale " which makes him say loftily that the propositions are not constant and that Arithmetic is self-contradictory. But you will see it very easily by the principles you have.‡

The remainder of the letter is concerned with proving that the difference of the cubes of any two consecutive natural numbers, when unity is subtracted, is six times the sum of all the numbers contained in the smaller one,§ and a statement of two geometrical problems. It should be noted that in spite of not hiavng the time to elucidate M. de Méré's difficulty Pascal still had time enough to write about these.

The reply from Fermat to Pascal's letter of July 29th, 1654, is missing, although we can infer its content. A letter from Fermat to Carcavi of August 9th, 1654 is worth considering first, however,

---

* Throw the " Sonnez ", i.e. throw 2 sixes. In playing backgammon if a player threw 2 sixes he would cry " Sonnez, le diable est mort ".

† Single die. Probability of one six = 1/6. Probability of no six = 5/6.

Probability of no sixes in n throws $\left(\dfrac{5}{6}\right)^n$

Probability of at least one six in n throws $1 - \left(\dfrac{5}{6}\right)^n = p$.

$n = 4$, $p = 671/1296 = 0.5177$

Two dice.   Probability of 2 sixes $= \dfrac{1}{36}$. Probability of not having 2 sixes

$= \dfrac{35}{36}$. Probability of no throw of 2 sixes in n thows $(35/36)^n$.

Probability of at least one throw of 2 sixes $= 1 - \left(\dfrac{35}{36}\right)^n = p$.

$n = 24$, $p = 0.4914$.
$n = 25$, $p = 0.5055^+$.

‡ The Chevalier de Méré was obviously such an assiduous gambler that he could distinguish empirically between a probability of 0.4914 and 0.5, i.e. a difference of 0.0086, comparable with that (0.0108) of the gambler who asked advice of Galileo.

§ $(n + 1)^3 - n^3 - 1 = 3n^2 + 3n = \dfrac{6n(n + 1)}{2}$.

since it throws considerable light on the modest, unassuming Fermat. This letter was possibly written before that of July 29th reached Fermat.

Fermat to Carcavi : *August 9th, 1654*

> I was delighted to have agreed with M. Pascal, for I value his talent highly and I believe him to be capable of solving any problem that he undertakes. The friendship he offers is so dear to me and so precious that I think it fair to make use of it in publishing an edition of my treatises. If this suggestion did not shock you, you could both help in bringing out that edition, of which I would allow that you should be the masters. You could clarify or augment what seems too brief and thus relieve me of a case which my work prevents me from undertaking. I would like this work to appear without my name, leaving altogether to you the choice of all the designations which could indicate the name of the author, whom you could qualify as just a friend.
>
> Here is the line which I have thought up for the second Part which will contain my inventions as to the numbers. It is a work which is still only an idea and which I may not have the leisure for putting fully on paper : but I will send a summary to M. Pascal of all my principles and first demonstrations, in which, I can promise you in advance, he will find everything not only new and up till now unknown but also astonishing. If you combine your work with his, everything will succeed and be completed in a short time, and we will yet be able to publish the first Part which you have in your power to do. If M. Pascal approves of my overtures, which are founded mainly on the great esteem which I have for his genius, his knowledge and his intellect, I will inform you of my numerical inventions.

This offer of Fermat's was never taken up, and one can only regard it as one of the greatest missed opportunities in the history of mathematics.

A short time before August 24th there must have been a letter from Fermat to Pascal setting out the answer to M. de Méré's difficulty, presumably in terms involving some sort of combinatory theory. The doubts which one has, possibly unjustly, about Pascal's understanding, deepen a little when we read his reply on August 24th to this missing letter. He says that Fermat's method is satisfactory when there are only two players but will not be applicable if there are more than two. Pascal then starts enumerating by exhausting the possibilities, supposing two players, the first, *A*, wanting two points to win and the second, *B*, wanting three. Since *A* needs 2 and *B* needs 3 points the game will be decided in four throws. The possibilities are :

AAAA	AAAB	AABB	ABBB	BBBB
	AABA	ABAB	BABB	
	ABAA	BAAB	BBAB	
	BAAA	ABBA	BBBA	
		BABA		
		BBAA		

In these enumerations every case where *A* has 2, 3, or 4 successes is a case favourable to *A* and every case where *B* has 3 or 4 represents a case favourable to *B*. There are 11 for *A* and 5 for *B*, so that the odds are 11 : 5 in favour of *A*. This is Pascal's own example, and because of what follows we note that he has included several cases where the game could have been determined in less than 4 throws.*

Pascal goes on to say that he has communicated Fermat's solution to Roberval, Professor of Mathematics at the Collège de France, member of the one-time Mersenne Academy and friend of Étienne Pascal. Roberval has the distinction of being the first known mathematician to raise the (invalid) objection, which objection has been repeated through the years—

What is mistaken is that the problem is worked out on the assumption that *four* games are played in view of the fact that when one man wins *two* games or the other wins *three*, there is no need to play *four* games : it could happen that they would play two or three or in truth perhaps four. And thus he could not see why you

---

* It is perhaps not necessary to point out that the method of solution is one which we would use today. Given that the game must finish in 4 throws then all possibilities for these 4 throws are equally likely. Out of these possibilities we pick out those favourable to *A*. But in order that each set of 4 shall be equally likely, all must be enumerated.

claim to find the fair division of stakes on the false assumption that
they play four games.*
(Pascal to Fermat.)

Pascal says he answered Roberval by saying that although it is
possible that the game may be finished in 2 or 3 throws, yet the
mathematician can suppose that the players agree to have four
trials, for

it is absolutely equal and immaterial to them both whether
they let the game take its natural course.

Leibnitz was perhaps referring to this point when he wrote about

the beautiful ideas about games of chance of Messieurs Fermat,
Pascal and Huygens which M. Roberval was not able or did not
wish to understand,

although what authority he had for the " did not wish " is not
stated.   Pascal continues his letter (to Fermat) by remarking
that

I must tell you that the solution for two players based on combinations is very accurate and true, but if there are more than two
players it will not always be correct.

He supposes three players A, B and C (of equal skill although this
is not stated). Let A want 1 point, B and C each 2 points to win,
so that the game must be finished in 3 trials, i.e. ABC, BBC (or A),
CCB (or A), etc. He writes down the following table in which
the 27 possibilities, correctly enumerated, are to be read
downwards

---

* AA BA BB Probability that A wins in 2 games $= \frac{1}{4}$.
  AB

AAA AAB ABB BBB Probability that A wins in 3 = Probability
  ABA BAB that he gets 1 point in 2 $\times$ Probability that he
  BAA BBA wins the third game $= \frac{1}{2} \cdot \frac{1}{2} = \frac{1}{4}$.
  Probability that A wins in 4 = Probability
  that he gets 1 point in 3 $\times$ Probability that
  he wins the fourth game $= \frac{3}{8} \cdot \frac{1}{2} = \frac{3}{16}$.
  $\frac{1}{4} + \frac{1}{4} + \frac{3}{16} = \frac{11}{16}$ as previously deduced.

AAA	AAA	AAA	BBB	BBB	BBB	CCC	CCC	CCC
AAA	BBB	CCC	AAA	BBB	CCC	AAA	BBB	CCC
ABC	ABC	ABC	ABC	ABC	ABC	ABC	ABC	ABC

1 1 1	1 1 1	1 1 1	1 1 1	1	1	1 1 1	1	1
	2			2	2 2 2	2		2
		3				3	3	3 3 3 3

It remained only for him to note the winners as

AAA   AAA   AAA   AAA   BBB   ABC   AAA   ABC   CCC

to give $A : B : C$ as 17 : 5 : 5, but he does not do this. He writes :

> The first man needs just *one* game to win : thus all the throws giving one A are favourable to him ; there are 19.
>
> The second man needs *two* games : thus all the throws giving two B's are his : there are 7.
>
> The third man needs *two* games : thus all the throws with two C's are his : there are 7.
>
> If from that one concluded that the stakes must be divided 19 : 7 : 7 one would make a gross error and I cannot believe you would do this. . . . We must add up the throws common to the two players as each being worth the whole amount but only worth half to each player. . . . Thus to share out the stakes we must multiply

> | | |
> |-----------------------------|-----|
> | 13 by one which makes | 13 |
> | 6 by one-half which makes | 3 |
> | 8 by zero which makes | 0 |
> | ─── | ─── |
> | 27 | 16 |
> | ─── | ─── |

and divide the sum of the products by the sum of the throws, 27, which gives the fraction 16/27.

He goes on by the same arguments to get the fractions $5\frac{1}{2}/27$ for B and C each, remarks that the method of solution is unfair and continues :

> The reason for this is that you make a false assumption, which is that *three* games are played invariably, instead of letting the play take its natural course, which is to play only until one man has got the number of games he needs, when play ceases.

This remark is surprising in the light of his reported remarks to Roberval, and it raises the uneasy suspicion that possibly M. de Roberval was being used as a stalking horse for Pascal's own conceptual difficulty. He goes on to say that if they agree to play 3 games then the ratios are $16 : 5\frac{1}{2} : 5\frac{1}{2}$, but if they break off when one of them has won then the ratios are 17 : 5 : 5 and that this can be shown by his general method which does not use combinations. It would seem perhaps that Pascal is in a muddle, even as he appears to have been in the missing first letter of the series, and that he did not really understand what he was doing.

Fermat writes briefly on August 29th, possibly in reply to this letter of Pascal's of August 24th—although this does not give much time for the mail—possibly in reply to an earlier letter of Pascal's which may be missing. He remarks shortly that the correct answer is 17 : 5 : 5 and that this comes out of the combinatorial method. He begins by speaking of the eleventh proposition* on the Arithmetic Triangle which Pascal has sent him, an episode which is a little puzzling. Fermat does not say what this proposition is, but writes as though he himself had also just

---

* It has already been noted that Pascal was not the originator of the Arithmetic Triangle and further that he must have known he was not. But the Eleventh Proposition of Pascal is just this:—

In the natural progression of numbers beginning with unity, any number whatever being added to the one immediately greater, produces the double of its triangle. The same number being added in the triangle immediately greater produces the triple of its pyramid. The same number added in the pyramid of the number immediately greater produces the quadruple of its triangulo-triangular, and so on to infinity by a method general and uniform.

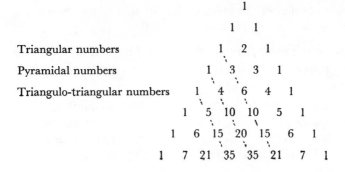

								1							
Triangular numbers						1	2	1							
Pyramidal numbers					1	3		3	1						
Triangulo-triangular numbers				1	4		6		4	1					

thought of it, while in the appendix to Pascal's note on the Arithmetic Triangle there is a note on the 11th corollary which reads as follows :

> This same proposition I have just expounded has come from the pen of that celebrated magistrate of Toulouse, M. de Fermat. And, what is admirable, without his having given me the least clue, not a word, he writes in the Provinces what I have invented in Paris, hour by hour, as our letters written and received at the same time witness. Happy am I to have coincided on this occasion, as I have done again in others in a way just as strange, with a man so outstanding in intellect and who in all his researches of the most sublime geometry is in the highest degree of excellence, so that his works that our long prayers have at last obtained from him will make him supreme among all the geometers of Europe.

Discounting the somewhat fulsome language there are two points here of importance. Fermat has not mentioned that he had already discovered the proposition and Pascal claims independent discovery of it. It would be in keeping with Fermat's modesty and carelessness about his own work perhaps, that he did not disclose that he had invented the proposition in 1636, some eighteen years before this present correspondence took place. The Abbé Mersenne stated the proposition and wrote to ask Fermat for a proof. Fermat generalised it, found a proof, and wrote a letter to the Abbé offering details of the proof both to him and to Sainte-Croix. Fermat himself applied it to the evaluation of parabolic areas. Roberval also occupied himself with the proposition, but his solution was not general and Fermat was not satisfied with it. It may be noted that Pascal, although still a boy, was a member of the Mersenne Academy all the time this correspondence was going on. A biographer of Pascal says :

> The ignorance of Pascal on this point leads us to think that although admitted as a child to the Mersenne Academy he did not follow their discussions with great regularity.

Perhaps it is better left at that. Fermat's letter concludes with the proposition that the squared powers of two, plus one, is always a prime number.

There does not seem to have been any reply by Pascal to this letter of Fermat's on August 29th, since Fermat writes again on September 25th obviously in answer to Pascal's letter of August 24th. He points out the error in Pascal's reasoning, shows that the answer is the same if 4 games (81 possibilities) are considered instead of 3, and gives an alternative proof of the 17 : 5 : 5 ratio. He does not take long over this and he then proceeds to state many of his conjectures in number theory. He adds in his simplicity :

> I am persuaded that since you know my way of proving this kind of proposition they will appear good to you and will give you room to make new discoveries. If I have time we will speak later about magic numbers and I will recall my old thoughts on this subject.

Pascal may have realised from this letter that he was getting outside his own mathematical comprehension. Moreover he was approaching the time of his second conversion when, except for a brief period, he gave up mathematics altogether. How much the feeling that he was of inferior intelligence, mathematically, to Fermat, contributed to his decision to give up mathematics it is impossible to say, but it was perhaps one of the factors which led him to lapse into the melancholia of his conversion. The last letter of the series, written by him to Fermat and dated October 27th, 1654, is strangely flat compared with the previous ones.

> Your last letter pleased me well. I admire your method for the problem of dividing the stakes, all the more so because I understand it perfectly : it is entirely original and has nothing in common with mine. It reaches the same conclusion very simply. Thus our mutual understanding is restored. But, Sir, even if I have agreed with you in this, look elsewhere for someone who would follow your numerical arrangements, the exposition of which you have so kindly sent me. I must confess these are far beyond my comprehension. I can but admire them and very humbly beg you to take the first opportunity of completing them. All our colleagues saw them last Saturday and valued them very highly : it is not easy to wait for such elegant and desirable things.

Pascal had had the intention of publishing a book on games of chance (*Aleae Geometria*) or so he wrote in a letter in 1654. Whether his difficulties in understanding Fermat or his conversion prevented this undertaking we do not know, but it did not come to completion. On the whole it would seem that Pascal's reputation as a mathematician is greater than he deserved, although there is no doubt that he was very competent. He was a boy of mathematical genius, a man of mathematical ability. How far he was pushed by his father, how far his father's insistence gave him the feeling that he must excel, it would be difficult to evaluate. But on several occasions Blaise seems to have appropriated as his own, mathematical ideas which belonged to other scientists, and the scientific world, possibly through Étienne's prestige and his loving imagination about his son's abilities, accepted them as original. In those days when all problems were discussed by all scientists it is difficult to understand how such misappropriation was allowed, but apparently it was. The feeling of Pascal's personality as it comes down the years through his letters and MSS is not a pleasant one. Fermat, on the other hand, seems to have been a man who even after 300 years attracts all who read about him.

There was one further exchange of letters between Fermat and Pascal in 1660. On July 25th, Fermat wrote from Toulouse to Pascal who was visiting Clermont-Ferrand, suggesting that they should meet at a place midway between the two towns. Fermat remarks there that his health is no better than Pascal's. Pascal replied on August 10th, saying that he appreciated the offer but that he was sick, and

> to talk freely with you about geometry is to me the very best intellectual exercise, but that at the same time I recognise it to be so useless that I can find little difference between a man who is only a geometrician and a clever craftsman. Although I call it the best craft in the world it is, after all, nothing else but a craft, and I have often said it is fine to try one's hand at it but not to devote all one's powers to it. . . . It is quite possible I shall never think of it again. . . . .

They never met, and there were no further letters. Pascal died in 1662 and Fermat in 1665.

chapter 10

Bills of Mortality

> Originally drawn up on long parchment rolls stored in the
> Treasury at Winchester, the survey was copied into two
> volumes christened by the English " Domesday " because
> there was no appeal against it. It was the most remarkable
> administrative document of the age : there is nothing like it
> in the contemporary annals of any other country. It en-
> abled the king to know the landed wealth of his entire realm,
> " how it was peopled and with what sort of men", what
> their rights were, and how much they were worth.
>
> ARTHUR BRYANT, *Makers of the Realm*

Modern statistical method and practice has its twin roots in the
calculus of probabilities and in the collection of numerical data.
It would be outside the main theme of this present treatise to
trace the rise of actuarial science and the growth of vital statistics,
but at this juncture, when the calculus of probabilities was well
launched, it is interesting to ask what the English were doing.
Pre-eminence in mathematics had passed from Italy to France
and was shortly to pass to England with the advent of Newton.
Speculations and research into probability theory did not,
apparently, interest mathematicians across the Channel. The
English, as usual, were preoccupied with concrete facts.

The idea of enumeration had been firmly installed in the
English consciousness by the activities of William the Norman*
who not only had carried out an exhaustive economic survey
but also instituted a check survey to find out whether the first
enumerators had done their work adequately. This idea of
enumeration of economic wealth and manpower is of course an

---

* To read " Domesday " is to become aware yet again of how much more
efficient the whole procedure could have been if the archivists had had some
idea of the Hindu system of writing down numbers.

old one : echoes of it are found in many ancient documents including the Book of Numbers. But although the enumeration of wealth was accepted, there seems to have been a taboo on speculations with regard to health, philosophers as late as in the eighteenth century implying that to count the sick or even the number of boys born was impious in that it probed the inscrutable purpose of God. This attitude of mind has persisted to a certain extent into our own times.

The parish registers in England began in 1538 as a consequence of one of the seventeen injunctions made during that year, nominally by Henry VIII but actually by Thomas Cromwell (1485–1540), his Grand Chancellor, who acted for him in matters connected with the Church. It was appointed that the parson, vicar or curate should keep a true and exact register of all weddings, christenings and burials. These registers, however, only tell one part of the story of a parish, since those who were of a different faith from the Church of England (for example Quakers*) and whose sect had their own burial grounds, would not appear in the burial register. Before the parish registers began, the first Bills of Mortality were printed. Bell in his *History of the Great Plague* states that the preparation of the Bills in time of plague originated from a request by the Privy Council to the Lord Mayor in October 1532 to furnish a return of plague deaths, so this may be one more debt that is owed to Thomas Cromwell. There are various other Bills during this century showing the number of plague deaths compared with all other sickness, but they were sporadic, and a regular weekly Bill " which contained an account of the christenings and burials as recorded by the parish clerks in London " did not make its appearance until December 21st, 1592. In 1594 this account was first made public, but it was discontinued on December 18th, 1595, after the plague epidemic ceased, and it did not begin again until December 29th in 1603, the first year of James I's reign. By 1625 these weekly bills of mortality had acquired repute, and the company of parish clerks obtained a decree from the Star Chamber for the setting up of

---

* Founded by George Fox (1624–1690).

a printing press in their hall " in order to the printing of the weekly and general bills within the City of London and the liberties thereof ". The Archbishop of Canterbury assigned a printer to do this, and one by one all the parishes of London, both within and without the walls, joined in the printing. At the start there were 97 parishes within the walls and 16 parishes without the walls. The bills were published weekly, with a general account for the whole year published on the Thursday before Christmas Day (see Fig. 6).

It would seem then that the vital statistician (the empirical probabilist) of 1660–1670 would have a reasonable amount of material on which to draw, but many of the bills are missing and many of the church registers were destroyed in the Great Fire of 1666. What is known about diseases causing deaths in that century comes chiefly from two sources, John Graunt's book *Natural and Political Observations on the Bills of Mortality* of 1662, and the complete collection of such material as existed up to 1668 plus all the yearly bills issued after that date compiled by an unknown* (?) writer of the eighteenth century. Graunt's work was of importance in that it gave the lead to other countries. For example, in a Paris journal of 1666 it is said that " to issue bills of mortality is a thing peculiar to the English", and the necessary alteration to enable a similar thing to be done in Paris was not added to the French legal code until 1667. The introduction into other European countries was after that date, sometimes much later.

The first English vital statistician seems to have been John Graunt with his book. Before considering the material with which he worked—the equivalent material for him that the results of the dice-throwers were for the continental mathematicians— such brief data as concern him should be recorded. Little is known of him except through Aubrey's *Brief Lives* and such as may be inferred from his own writings. Graunt was a citizen of

---

* I have the feeling that Major Greenwood succeeded in identifying this author, but I cannot find the reference.

Fig. 6 (*opposite*)—**Bill of Mortality recorded by London parish clerks for the year 1665**

A generall Bill for this present year, ending the 19 of December 1665. according to the Report made to the KINGS most Excellent Majesty.

By the Company of Parish Clerks of London, &c.

Parish	Buried	Plag	Parish	Buried	Plag	Parish	Buried	Plag	Parish	Buried	Plag
St A'bans Woodstreet	200	121	St Clements Eastcheap	28	20	St Margaret Moses	58	25	St Michael Cornhill	104	52
St Alhallowes Barking	514	330	St Dionis Back-church	78	27	St Margaret Newfisht.	114	66	St Michael Crookedla	179	133
St Alkallowes Breadst	35	16	St Dunstans East	665	150	St Margaret Pattons	49	24	St Michael Queenhit	103	122
St Alhallowes Great	455	426	St Edmunds Lumbard	170	30	St Mary Abchurch	99	54	St Michael Que oe	44	18
St Alhallowes Honila	10	5	St Ethelborough	195	106	St Mary Aldermanbury	181	109	St Michael Royall	152	116
St Alhalowes Lesse	239	175	St Faiths	104	70	St Mary Aldermary	105	75	St Michael Woodstreet	122	62
St Alball. Lumbardstr	90	52	St Fosters	128	105	St Mary le Bow	64	26	St Mildred Breadstreet	59	26
St Alhallowes Staining	185	112	St Gabriel Fen-church	69	39	St Mary Bothaw	35	20	St Mildred Poultrey	68	46
St Alhallowes the Wall	500	356	St George Botolphlane	41	17	St Mary Colechurch	17	6	St Nicholas Acons	46	28
St Alphage	271	115	St Gregories by Pauls	376	232	St Mary Hill	94	64	St Nicholas Coleabby	125	91
St Andrew Hubbard	71	25	St Hellens	108	75	St Mary Mounthaw	56	37	St Nicholas Olaves	90	68
St Andrew Undershaft	274	189	St James Dukes place	162	190	St Mary Summerset	342	262	St Olaves Hartstreet	237	160
St Andrew Wardrobe	476	302	St James Garlickhithe	189	116	St Mary Stayning	47	27	St Olaves Jewry	54	32
St Anne Aldersgate	282	197	St John Baptist	138	103	St Mary Woolchurch	65	33	St Olaves Silverstreet	250	132
St Anne Blacke-Friers	652	467	St John Euangelist	9		St Mary Woolnoth	75	38	St Pancras Sopertane	30	13
St Antholins Parish	58	33	St John Zacharie	85	54	St Martins Iremonger	21	11	St Peters Cheape	61	35
St Austins Parish	43	20	St Katherine Coleman	299	213	St Martins Ludgate	195	128	St Peters Cornehill	136	76
St Barthol. Exchange	73	51	St Katherine Creechu.	235	231	St Martins Orgars	110	71	St Peters Pauls Wharfe	114	86
St Bennet Fynch	47	22	St Lawrence Jewry	94	48	St Martins Outwich	60	24	St Peters Poore	79	47
St Benn. Grace-church	57	41	St Lawrence Pountney	214	140	St Martins Vintrey	417	349	St Stevens Colemanst	360	391
St Bennet Pauls Wharf	355	172	St Leonard Eastcheap	42	27	St Matthew Fridaystt.	24	6	St Stevens Walbrooke	34	17
St Bennet Sherehog	11		St Leonard Fosterlane	335	255	St Magdalen Milkstreet	44	23	St Swithin	93	
St Botolph Billingsgate	83	50	St Magnus Parish	103	60	St Magdalen Oldfishstr.	176	121	St Thomas Apostle	63	110
Christs Church	653	457	St Margaret Lothbury	100	66	St Michael Bassishaw	253	164	St Trinitie Parish	115	79
St Christophers	60	47									

Buried in the 97 Parishes within the walls, — 15207 whereof of the Plague — 9887

Parish	Buried	Plag	Parish	Buried	Plag	Parish	Buried	Plag
St Andrew Holborne	3958	3103	Bridewell Precinct	230	179	St Dunstans West. — 958		
St Bartholmew Great	493	344	St Botolph Aldersga.	997	755	St George Southwark	1613	1260
St Bartholmew Lesse	193	139	St Botolph Algate	4926	4051	St Giles Cripplegate	8069	4838
St Bridget	2111	1427	St Botolph Bishopsg.	3464	2500	St Olaves Southwark	4793	2785

St Saviours Southwark 3271 2446
St Sepulchres Parish 4509 2746
St Thomas Southwark 475 371
Trinity Minories 168 123

Buried in the 16 Parishes without the walls — 41351 Whereof, of the Plague — 28888 At the Pesthouse 159 156

Parish	Buried	Plag	Parish	Buried	Plag	Parish	Buried	Plag
St Giles in the Fields	4457	3216	St Katherines Tower	956	601	St Magdalen Bermon.	1943	1262
Hackney Parish	232	132	Lambeth Parish	798	537	St Mary Newington	1272	1004
St James Clarkenwell	1863	1377	St Leonard Shordisch	2669	1949	St Mary Islington	696	593

St Mary Whitechappell 4766 3855
Redriffe Parish 304 210
Stepney Parish 8598 6583

Buried in the 12 out-Parishes, in Middlesex and Surrey — 18554 Whereof, of the Plague — 21420

Parish	Buried	Plag
St Clement Danes	1969	1319
St Paul Covent Garden	408	261
St Martins in the Fields	4804	2883
St Mary Savoy — 303 198		
St Margaret Westminst.	4710	3742
buried at the Pesthouse — 156		

Buried in the 5 Parishes in the City and Liberties of Westminster — 12194 Whereof, of the Plague — 8403

The Total of all the Christnings — 9967
The Total of all the Burials this year — 97306
Whereof, of the Plague — 68596

The Diseases and Casualties this year.

Disease	Number	Disease	Number	Disease	Number
Abortive and Stilborne	617	Executed	21	Palsie	30
Aged	1545	Flox and Small Pox	655	Plague	68596
Ague and Feaver	5257	Found dead in streets, fields, &c.	20	Plannet	6
Appoplex and Suddenly	116	French Pox	86	Plurisie	15
Bedrid	10	Frighted	23	Poysoned	1
Blasted	5	Gout and Sciatica	27	Quinsie	35
Bleeding	16	Grief	46	Ruckets	557
Bloody Flux, Scowring & Flux	185	Griping in the Guts	1288	Rising of the Lights	397
Burnt and Scalded	8	Hangd & made away themselves	7	Rupture	34
Calenture	3	Headmouldshot & Mouldfallen	14	Scurvy	105
Cancer, Gangrene and Fistula	56	Jaundies	110	Shingles and Swine pox	2
Canker, and Thrush	111	Imposthume	227	Sores, Ulcers, broken and bruised Limbs	82
Childbed	625	Kild by severall accidents	46	Spleen	14
Chrisomes and Infants	1258	Kings Evill	86	Spotted Feaver and Purples	1929
Cold and Cough	68	Leprosie	2	Stopping of the stomack	332
Collick and Winde	134	Lethargy	14	Stone and Strangury	98
Consumption and Tissick	4808	Livergrown	20	Surfet	1251
Convulsion and Mother	2036	Meagrom and Headach	12	Teeth and Worms	2614
Distracted	5	Measles	7	Vomiting	51
Dropsie and Timpany	1478	Murthered and Shot	5	Wenn	8
Drowned	50	Overlaid & Starved	45		

	Males	5114		Males	48569	
Christned	Females	4853	Buried	Females	48737	Of the Plague 68596
	In all	9967		In all	97306	

Increased in the Burials in the 130 Parishes and at the Pest-house this year — 79009
Increased of the Plague in the 130 Parishes and at the Pest-house this year — 68550

London and a tradesman. He was a haberdasher of small wares at the Seven Stars in Birchin Lane in the City, a freeman of the Drapers' Livery Company, a captain of the Train-Bands of the City—a Commonwealth man. It is not possible to deduce whether this political application was through force of circumstance—the City of London was generally against the King—or through conviction. The somewhat fulsome references to Charles II in his book read like someone trying to convince his world that his change of heart is genuine. He comes into the public eye in 1662 with the publication of this book. Why he wrote it is not certain. He wrote of his fellow merchants that

> most of them who constantly took in the weekly bills of mortality, made little or no use of them than so as they might take the same as a text to talk upon in the next company : and withal, in the plague time, how the sickness increased or decreased, that so the rich might guess of the necessity of their removal, and tradesmen might conjecture what doings they were like to have in their respective dealings.

Perhaps he originally started the investigation with a view to predicting trade, or perhaps he just wanted to know.

John Graunt's manuscript was presented to the newly formed Royal Society in 1661 and he was elected a Fellow in 1662, presumably on the strength of it. Stories (probably apocryphal) are told that difficulties were raised about his election because he was just a shopkeeper, and that Charles II recommended him personally to the Society, adding moreover that if they found any more such shopkeepers they should also admit them. Perhaps the fulsome references were not lost on the king, although this would imply that he had read the book. It may have been through his Fellowship of the Royal Society that Graunt met Pepys. An extract from the famous diary (April 26th, 1668) runs :

> To the Crown Tavern and thither I and stayed a minute leaving Captain Graunt telling pretty stories of people that have killed themselves or have been accessory to it in revenge on other people, or to mischief other people.

Graunt's name disappears from the books of the Royal Society after 1666. He was in some sort of trouble after this time and there are suggestions that he was ruined by the Great Fire. His last years were undoubtedly clouded. Some time in this last period he changed his religion and became a Roman Catholic. He died in 1674 and (Aubrey writes) " his death is lamented by all good men who had the happiness to know him".

It would be agreed that the risks inherent in the processes of birth, marriage and death can only be calculated for a short forward period, and then only provided the structure of the population at risk is not changing. Graunt had some idea that the structure of the population of London *was* changing but, not surprisingly, he did not know what to do about it, and so he fell back on the doctrine of the stability of statistical ratios. He was not the first to propound this doctrine, but it is an obvious way of analysing such material and he probably came to it by himself. The earlier workers in the demographic field—and their successors over several centuries—believed that this stability was a direct manifestation of the purpose of God. For instance, Florence Nightingale,* after some lengthy calculations, wrote :

> The true foundation of theology is to ascertain the character of God. It is by the aid of Statistics that law in the social sphere can be ascertained and codified, and certain aspects of the character of God thereby revealed. The study of statistics is thus a religious service.

Her attitude is the inverse of that which Graunt possibly had to face, since he would almost certainly have been accused of impiety for his investigations.

Graunt, then, was the first Englishman to calculate empirical probabilities on any scale, and because his work was the fore-runner of so much it is perhaps worth while to look at the material on which he did his calculations. There seem to have been many flaws in the data, any one of which must have been enough to jeopardise the stability of the ratios. First, the weekly Bills were

---

* With truth indeed did she call herself "a passionate statistician".

based on returns made by each parish clerk to the Hall of the company of parish clerks. The clerks were not too scrupulous about this, some weeks not making a return and then lumping together all the figures in the week in which they did make it. Second, the Bills only relate to baptisms which were administered according to the rites of the Church of England, and besides those baptised in other faiths there were many poor people who did not ever have their children baptised. Third, only those buried according to Church of England rites were counted.* Fourth, persons who died in the parish but who were carried elsewhere for burial would not appear in the total. Fifth, only the parochial cemeteries were included, and many other burial grounds belonging to the Church of England, such as St. Paul's Cathedral, Westminster Abbey, the Temple Church, Lincoln's Inn Chapel, Charterhouse, and so on did not report their burials.

In the reproduction of the Bill of 1665 it will be noticed that deaths are divided by cause. Since death certificates, on which the cause of death supposed by the medicals is recorded, were not instituted until 1836, how were these figures obtained? The procedure appears to have been the following. Every parish appointed a searcher, an ancient matron whose business it was to examine the corpse and report on the distemper. The searchers were deliberately chosen to be persons of low intellectual calibre ; nothing more was asked of them than that they should relate what they heard.

> For the wisest person in the parish would be able to find out very few distempers from a bare inspection of the dead body and could only bring back such an account as the family and friends of the deceased would be pleased to give. And it should always be remembered from whom this account comes, that proper allowances may be made for mistakes and misrepresentations in those distempers which they might not be very able to know or not very willing to own.

---

* John Tillotson, Dean of St. Paul's, estimated that during one of the plagues the Quakers buried over one thousand a week for several weeks, and these would not be recorded.

Besides the low intellectual calibre of the searchers, uncertainty of diagnosis was another large source of error. John Tillotson, Dean of St. Paul's, notes this, and an eighteenth-century commentator remarks that both spotted fever deaths and plague deaths appear to increase together, suggesting that " there is reason therefore to suspect that this fever was the same from the beginning as the true plague".

In order to calculate empirical probabilities and to compare one disease with another and one year with another it is necessary to have some idea of the population at risk, and of this Graunt was well aware. It was a matter of some discussion among demographers of the seventeenth century as to which had been the worst plague year. Graunt decided to take the number of christenings of the population at risk and produces the following figures, adding the year 1665 later :

Period	All Burials	Plague Victims	% Plague	Christened	(Christened/ Plague) %
1592	26,490	11,503	43	4,277	37
1593	17,844	10,662	60	4,021	38
1603	42,042	36,269	86	4,784	13
1625	54,265	35,417	65	6,983	20
1636	23,359	10,400	45	9,522	92
1665	97,306	68,396	70	9,967	15

From this he concludes that 1603 was the worst year. Such evidence as there is suggests that the christenings dropped markedly in a plague year, but whether this was because people fled the city or because they were afraid at a time of mortality to take their children to church to be christened one cannot say. That the effect of the fearful mortality from the plague was not long-lasting is indicated by the fact that the year after the plague year christenings appear to be back to normal.

The fact that the plague deaths reached a maximum in weekly deaths in the late summer and early autumn was one which the early demographers could not fail to notice, and they were preoccupied with it. The years were classified according to the number of plague deaths ; if the number of these was greater than 200 in any particular year, and the yearly deaths exceeded those of the preceding year, that year was called " sickly". Since the population of London was increasing rapidly all the time it follows that an increasing number of years would be sickly. Graunt classifies 1618, '20, '23, '24, '32, '33, '34, '49, '52, '54, '56, '58, '61 as sickly years and adds (1661/2) :

> As to this year 1660, though we would not be thought super-stitious, yet it is not to be neglected, that in the said year was the King's restoration to his empire over the three nations, as if God Almighty had caused the healthfulness and fruitfulness thereof to repair the bloodshed and calamities suffered in his absence. I say, this conceit doth abundantly counterpoise the opinion of those who think that plagues come in with kings' reigns, because it happened so twice, viz. anno 1603 and 1625 ; whereas as well as the year 1649, wherein the present king commenced his right to reign, as also the year 1660, wherein he commenced the exercise of the same, were both eminently healthful : which clears both monarchy and our present king's family, from what seditious men have surmised against them.*

Graunt's importance both as a statistician and an empirical probabilist lies possibly in his attempts to enumerate the fundamental probability set, the population of London at risk to the several diseases such as are given in the Bills of Mortality. He points out first that each country marriage produces four children and then constructs tables which show that the male population of London was continually augmented by immigration from the country districts. This would mean among other things, that the structure of the population of London was continually changing, probably by immigration into the 20–30 male age-group, with the young men from the country tramping into London to seek their fortune. There would be marked fluctu-

---

* Graunt, described as a Roundhead, was presumably trimming his sails to catch the prevailing wind. It is the English habit and would undoubtedly have been prudent in those troubled times. One wonders at his thoughts in 1665.

ations due to the Civil War and the plague years, but there is no doubt that this was a time of great growth for London.

Data whereby inferences may be drawn, or even plausible guesses made, are very scanty. In the census of 1631 there were, men, women and children, in the liberties* of London, 130,178. This number an eighteenth-century chronicler thinks is too small, since

> I never observed more enormous mistakes in any matter than concerning the number of people, ale-houses, coaches, ships, seamen, watermen . . . .

From a consideration of the growth of parishes within and without the walls this chronicler estimates the population in 1661 as 403,000. Graunt proceeded in *his* estimation in several ways, including a sample survey. He remarked that he was put off for a long time from trying to estimate the number in the populace by the misunderstood example of the biblical David. However, eventually he satisfied his conscience and having also found out that there was nothing against the law in what he proposed to do, method 1 was as follows. He said that the number of child-bearing women was about twice the number of registered births. " Forasmuch as such women, one with another, have scarce more than one child in two years." He also remarked that the average yearly burials about the 1660 period were of the order of 13,000. The number of births was always less than the number of burials, so he put these at 12,000. Consequently the number of " teeming women " was 24,000. There might be twice as many women aged 16–76 as between 16 and 40, and about 8 persons (man, wife, 3 children, 3 servants or lodgers) per family, so that $8 \times 2 \times 24,000 = 384,000$ persons.

As a check on this figure we have method 2—" by telling the number of families in some parishes within the walls." How he did this " telling " he did not relate, not did he explain his result

---

* The liberties of London were circumscribed by Temple Bar, Holborn Bar, Smithfield Bars, Shoreditch Bars, Whitechapel Bars, the Tower, and the Meat Market at Southwark.

that 3 out of 11 families die per annum.* Since 13,000 persons died, we have 11. 13,000/3 families in all, which is 48,000 families as before. For method 3 he took the 1658 map of London, guessed that in 100 square yards there might be 54 families (i.e. about 209 persons per acre), supposing that each house has a 20-foot frontage. This gave him 11,880 families within the walls. Since 3,200 persons died per year within the walls and 13,000 altogether, it followed that the housing within the walls was about 1/4 of that outside it, and consequently the estimate was that there were 4 × 11,880 = 47,520 families in London.† Graunt in his arguments in other parts of his treatise assumed what is equivalent to an equal death-rate over 20 years and a uniform population density over London. It is curious, therefore, that he did not use the same arguments here and also make use of the 1631 census in which the population is given as c. 130,000 persons. The yearly burials for 1631, 1632 and 1633 are not complete, but those for 1634 and 1635 are 15,625 which, considering the degree of approximation involved, is close enough for this present purpose. The yearly burials for 1663 and 1664—in neither year are there as many as 10 reported having died from plague—are 15,326 + 18,297 = 33,623 or just about twice what they were in 1634 and 1635. Accordingly we might estimate the population exposed to risk in 1665 in the area covered by the bills of mortality as c. 260,000. This is less than Graunt and much less than those of his contemporaries who put the figure at millions. The effect of the fluctuating immigration figure cannot be allowed for, but one would suggest that it was unlikely to be large enough to increase this figure by as much as 50 per cent. Proceeding in this way, the estimated deaths from plague per 1000 persons at risk are 500 for 1603, 272 for 1625, 80 for 1636 and 264 for 1665, thus bearing out Graunt's analysis that 1603 was the " worst " plague year. Thus one in four of the estimated population at risk in 1665 died of plague, and since this population

---

* I am a little puzzled by this statement since, except in a plague year, families would not die out as units.

† I may perhaps be pardoned for the remark that it smacks of the procedure of think of a number and work round to it in as many ways as you can.

undoubtedly decreased rapidly as the plague deaths increased the actual rate would be much higher.

There is a great deal more to Graunt's treatise than I have mentioned briefly here. He discussed the sex ratio, the general mortality from different kinds of diseases, and the number of men who could bear arms, among many other topics. But enough has been said to indicate the importance of his work and the inadequate material from which he tried to draw inferences. He gave impetus to the collection of vital statistics, to life-tables, to insurance ; he illustrates as none other of his time and race could that the empirical approach of the English to probability was not through the gaming table but through the raw material of experience.

References

Apart from Aubrey's *Brief Life* and the Preface to

    A Collection of the Yearly Bills of Mortality (1657–1758)

information about Graunt and his work is rather scanty. These two sources were obviously used by

    Major GREENWOOD, "Graunt & Petty," *J. Roy. Statist. Soc.*, **91,** and "Graunt & Petty—a Re-statement," *J. Roy. Statist. Soc.*, **96,**

and I have perforce used them in my turn. For background material the two books by Bell,

    E. T. BELL, *The Great Plague* ; *The Fire of London*

are illuminating and accurate. There is much research work which might be done here, particularly in attempting to estimate the population of London at risk to the plague.

chapter 11

Christianus Huygens

No man is born into the world whose work is not born with
him. There is always work, and tools to work with, for those
who will.

JAMES RUSSELL LOWELL

If one looks on the first calculations of a probability by Cardan
and by Galileo as unimportant and says that the real begetter of
the calculus of probabilities is he who first put it on a sound footing,
then we should pass over not only Cardan and Galileo but also
Pascal and Fermat. For although Fermat, if provoked, could have
done as much and more than his successors, yet the fact remains
that his contribution was in effect the extension of the idea of the
exhaustive enumeration of the fundamental probability set, which
had already been given by Galileo. The scientist who first put
forward in a systematic way the new propositions evoked by the
problems set to Pascal and Fermat, who gave the rules and who
first made definitive the idea of mathematical expectation was
Christianus Huygens, Lord of Zelem and of Zuylichem. Huygens
was born at The Hague on April 14th, 1629, and died in the
Netherlands on July 8th, 1695. The place of his death is surmised
to be The Hague but this is not certain. He was buried there in
the St. Jacobkerk on July 17th, 1695. His father, also Christianus,
was Secretary to the Council of State of the Republic of the
United Provinces, and his mother was a member of one of the
first families in Holland. His elder brother, Constantin, was for a
time private secretary to Frederick, Prince of Orange. Christianus
was thus born into a family of wealth and position and was
expected, as a young man, to study in order to fit himself for that

110

position, and to travel abroad to meet influential persons and to obtain a cultural background.

It is said that he was first taught classics and mathematics by his father. His first tutor, at the age of 15, was a mathematician from Amsterdam called Stampisen, of whom Descartes had no great opinion. In 1645 (aetat. 16) he went to the University of Leyden to study law under Vinnius, but he still went on working at mathematics and transferred in 1646 to the newly established University of Breda for two years. He presumably went there to study under Francis van Schooten and another mathematician called J. Pell. It was said that during these two years he made rapid strides and attracted the attention of Descartes, although they never met. Huygens wrote *Theorems on the quadrature of hyperbolae, ellipses and circles* (Leyden, 1647) when he was 18, and *New inventions concerning the magnitude of the circle* (Leyden, 1654) when he was 25. These are the efforts of a young man and are of little interest. Perhaps he owed his noble birth for their publication. In 1655 he visited France for the first time to receive a doctorate in law from the Protestant University at Angers ; either on the way there, or on the way back, he stopped in Paris, for we are told that he arrived there in July of that year accompanied by his brother Ludovick and his cousin, Doublet. He stayed in Paris until the end of November.

Pascal was then enduring the miseries of his second conversion and Huygens did not see him. Neither did he see Fermat, who was still in Toulouse, nor Carcavi, but he was friendly with Mylon, one of Carcavi's friends, and at least acquainted with Roberval.* It is not surprising, therefore, that Huygens came to hear of the existence of the problem of points. He was not, however, told of the solutions of Fermat and of Pascal nor of the methods which they followed, which may have been because his Parisian friends were not competent to explain them. He was obviously intrigued with the problem, since on his return to

---

* It was this same Roberval to whom the Chevalier de Méré addressed his puzzles (as well as to Pascal) and who got involved in the wrong enumeration of the fundamental probability set.

Holland he set himself to work.   In March, 1656, he wrote to
van Schooten that he had written on dice-games and that he
would send the manuscript to him.   On April 18th he wrote to
Roberval and said that van Schooten thought the manuscript
should be printed.   Huygens posed to Roberval a problem which
is the equivalent of the Proposition 14 which he subsequently
published, but we are not told whether Roberval solved it or not.
The proposition was sent to Roberval because, Huygens said, he
wanted to see whether Roberval would reach the same solution
as himself.   The manuscript on dice-games, in publishable form,
was only sent to van Schooten and even so must have been without
the Exercises : for the Exercises owe their genesis to a three-party
correspondence between Huygens, Carcavi and Mylon which
did not begin until the end of April.

It was agreed between Huygens and van Schooten that
Huygens' treatise should become the fifth book of the *Exercitationes mathematicae* which van Schooten was even then preparing
to publish.   These Mathematical Exercises were to appear in two
editions, one printed in Latin and the other in Dutch.   It is said
that Huygens wrote his treatise in Dutch and experienced some
difficulty with the translation into Latin because of the necessity
of inventing technical terms.*   Possibly his translation was not
very good since van Schooten, after receiving it, replied by sending
Huygens another translation which he had done himself and which
was the one that he finally published.   While waiting for the
publication of his treatise, which did not appear until the following
year (1657), Huygens became more and more anxious to know
whether his solutions and his methods agreed with those of
the French mathematicians.   Not having received any reply to his
letter to Roberval of April 18th, 1656, he wrote to Mylon re-
peating the problem he had put to Roberval and adding others.

---

* On the whole this seems a little unlikely, but a parallel may be noted.
James Bernoulli's note-books, which are housed in the University Library
at Basel, have the orthodox mathematics written in Latin. When Bernoulli
tackles a probability problem all his arguments are written in French. Yet
there are very few words used which would not have appeared in the
classical texts.

The solutions which Mylon sent him, which were partly wrong, did not help him very much, but at the same time Carcavi, through Mylon, sent him the main outlines of the problem of points as discussed between Fermat and Pascal. More, on June 22nd, 1656, Carcavi sent to Huygens Fermat's solution to the problem, which Huygens found to be the same as he had worked out for himself. More still, Fermat posed to Huygens other and more difficult problems which Huygens eventually used for his *Exercises.* It is related that on the same afternoon that Huygens received these problems he found the solution of them all and that he replied immediately. It is clear Huygens was very quick in finding the answers since he gave the solutions in a letter to Carcavi on July 6th, 1656, which letter was passed on to Mylon, to Fermat and to Pascal.

Huygens now settled down to await with impatience the reply to his letter. This impatience can be felt when it is noted that he wrote to Roberval on July 27th and asked to be informed why neither Mylon nor Carcavi had answered him. Carcavi eventually did answer but not until the beginning of October. This time he told Huygens that he had used the same proposition as had Pascal, that he (Carcavi) did not see how it applied to the problem of points, and he added (wrongly) about this problem that Pascal had only found the rule for one player needing one point or two points. This led Huygens to take up the problem of points again and he wrote first an appendix to his treatise and then his Proposition IX* which he caused to be inserted in the text. The setting up in type was slow. Van Schooten wrote to Huygens in March 1657 that it would be ready in a few weeks, but it was actually August before it appeared under the title *De Ratiociniis in Aleae Ludo.* There are two forewords. The first, from van Schooten, runs :

> *To the Reader :* When I had already decided to close my book of Exercises, it occurred to me that there were still more remarkable and amusing subjects left out. If I had added them to my sections and if I had succeeded in treating them worthily, they would have

---

* See our page 116. Proposition IX is very like that which Pascal wrote to Fermat.

greatly distinguished my work and perhaps made the exercises easy and more profitable. However, the difficulty that I would have had in developing them and the work that this would have entailed, seemed to me to be too much. This is why—even though among other things treated in the preceding sections of this work I have discussed some of the most beautiful and subtle theorems set out by earlier mathematicians and by the best mathematicians of this century—it has not seemed inopportune to me, in order to extend the applications of the art of mathematics, to add here, instead of my further work, the theorems which have been invented lately on the calculus of games of chance by the Noble and Distinguished Lord, Christian Huygens. I presume that his wit will please you, Reader, better than the considerations of the author : above all, he has used the same analysis which I have used, and of which I have taught him the elements,* and so he has set out for those who would study this subject a method for working out similar problems. If in this way I have given you, dear Reader, out of the rest of my work, enough topics to exercise yourself in the study of analytical geometry you will understand, I hope, my good will towards you. Adieu.

The second foreword was by Huygens himself, and he wrote :

To M. Franciscus van Schooten. Sir, knowing that in publishing the valuable fruit of your intelligence and zeal you are trying to make others realise, by the diversity of the subjects treated, the great width of the field over which our marvellous Algebraic Art extends, I do not doubt that the present writing on the subject of the calculus of games of chance will serve towards that end. In fact, the more it seems difficult to determine by reason what is uncertain in chance, the more the Algebra which determines this result appears admirable. It was at your request and in response to your repeated demands† that I began to write about this calculus of chance that you have judged worthy to be set among the results of your own researches. Not only do I willingly give you permission to publish it in this way, but I am sure that this method of publication will be entirely to my advantage. For if some readers think that I have worked on subjects of small im-

---

* It is not clear what this means. The theory of arithmetical proportions ?
† It does not seem to have been quite like this.

portance, nevertheless they will not condemn as having no utility and not worth any purchase price, what you have adopted in this way as though it was your own work, after having translated it, not without trouble, from our tongue into Latin. I would like to believe that in considering these matters closely, the reader will soon understand that I do not treat here a simple game of chance but that I have thrown out the elements of a new theory, both deep and interesting. The Problems leading to this theory are not, it seems to me, to be judged any easier than those of Diophantus, but one will find them more amusing perhaps in that they deal with the simple properties of numbers. It should be said, also, that for some time some of the best mathematicians of France have occupied themselves with this kind of calculus so that no one should attribute to me the honour of the first invention. This does not belong to me. But these savants, although they put each other to the test by proposing to each other many questions difficult to solve, have hidden their methods. I have had therefore to examine and to go deeply for myself into this matter by beginning with the elements, and it is impossible for me for this reason to affirm that I have even started from the same principle. But finally I have found that my answers in many cases do not differ from theirs. You will find at the end of this treatise that I have proposed some questions of the same kind without indicating the method of proof. This is in the first place because I think it would cost me too much work to expand succinctly the reasoning leading to the answer, and in the second place because it seems useful to me to leave something for my readers to think about (if I have any readers) and this will serve them both as an exercise and a way of passing the time : Your devoted servant, Christian Huygens de Zuylichem.

This treatise of Huygens *De Ratiociniis in Aleae Ludo* was, it is said, warmly received by contemporary mathematicians, and for nearly half a century it was the unique introduction to the theory of probability. It was not until the tremendous researches of the 1690–1710 period which resulted in *Essai d'Analyse sur les Jeux de Hasard* (Montmort, 1708), *Ars Conjectandi* (James Bernoulli, 1713), *Calcul des Chances, à la statistique générale, à la statistique des décès et aux rentes viagères* (Nicholas Struyck, 1713), and *Doctrine of Chances* (Abraham de Moivre, 1718) that Huygens' work was superseded, and then not entirely since James Bernoulli incorporated it in the

Ars Conjectandi. Two English translations of it appeared during this period. In 1692 there is an anonymous work *Of the Laws of Chance* attributed by Todhunter to John Arbuthnot, and in 1714 a treatise by W. Browne entitled *The value of all chances in games of fortune, cards, dice wagers, lotteries, etc. mathematically demonstrated.*

Huygens' fourteen propositions run as follows :

I :   To have equal chances of getting a and b is worth $(a + b)/2$.

II :   To have equal chances of getting a, b or c is worth $(a + b + c)/3$.

III :   To have p chances of obtaining a and q of obtaining b, chances being equal, is worth $(pa + qb)/p + q$.

IV :   Suppose I play against an opponent as to who will win the first three games and that I have already won two and he one. I want to know what proportion of the stakes is due to me if we decide not to play the remaining games.

V :   Suppose that I lack one point and my opponent three. What proportion of the stakes, etc.

VI :   Suppose that I lack two points and my opponent three, etc.

VII :   Suppose that I lack two points and my opponent four, etc.

VIII :   Suppose now that three people play together and that the first and second lack one point each and the third two points.

IX :   In order to calculate the proportion of stakes due to each of a given number of players who are each given numbers of points short, it is necessary, to begin with, to consider what is owing to each in turn in the case where each might have won the succeeding game.

X :   To find how many times one may wager to throw a six with one die.

XI :   To find how many times one should wager to throw 2 sixes with 2 dice.

XII :   To find the number of dice with which one may wager to throw 2 sixes at the first throw.

XIII :   On the hypothesis that I play a throw of 2 dice against an opponent with the rule that if the sum is 7 points I will have won but that if the sum is 10 he will have won, and that we split the stakes in equal parts if there is any other sum, find the expectation of each of us.

XIV :   If another player and I throw turn and turn about with

2 dice on condition that I will have won when I have
thrown 7 points and he will have won when he has thrown
6, if I let him throw first find the ratio of my chance to his.

His proof of Proposition X is interesting because he wraps it up in
expected values.

It is certain that the gambler who wagers to throw a 6 in a
single throw has one chance of winning and 5 of losing. For he has
five throws against him and only one for him. Call the stakes a.
He has therefore one chance of getting a and 5 chances of not getting
it, so that by the second* Proposition he will expect $a/6$. There
remains $5a/6$ for his opponent. He who plays a single game of one
throw therefore can only count odds 1 against 5.

The proportion due to the gambler who wagers to throw a
6 in two throws, is calculated in the following way. If he throws a 6
the first time, he wins a. If he fails at the first throw, he still has
another which is worth $a/6$ to him by the previous argument. But
he has only one chance of throwing a 6 at the first throw and he has
5 chances of not getting it. He has therefore at the beginning one
chance of obtaining a and 5 chances of obtaining $a/6$, which is
worth $11.a/36$ by the second proposition. There remains $25.a/36$
for his opponent. He who throws and wagers for a six in 2 throws
has odds therefore of 11 : 25 which is less than 1 : 2.

He proceeds in this way to argue that the odds for throwing a six
in three throws is 91 : 125, for four 671 : 1296, for five 4651 : 7776,
for six throws 31031 : 46656. It is difficult to see why he finds it
necessary to introduce a into the argument. Proposition XIV
is wrapped up in the same way. He writes :

Let x be the value of my chance, and let a be the stakes. The
value of my opponent's chance is therefore $a - x$. It is evident
also that each time it is his turn to throw, my chance will have the
value x. But each time that it is my turn to throw, my chance will
have a greater value, say y. Now among the 36 throws it is possible
to make with 2 dice, there are 5 which will give 6 points to my
opponent and will let him win and 31 throws to his disadvantage,
that is to say which will lead to my turn to throw. I have 5 chances

---

* He means Proposition III.

of having 0 when he throws the first time and 31 chances of having y, which, following the third proposition, is worth $31y/36$ to me. But we have supposed

$$\frac{31y}{36} = x \quad \text{giving} \quad y = \frac{36x}{31}.$$

We have supposed also that my chance is worth y when it is my turn to throw. But when I throw I have 6 chances of winning a, since there are 6 throws of 7 points which will make me win ; and I have 30 chances of giving the turn back to my opponent, that is to say of having for my part, x. The value y is therefore equivalent to 6 chances of having a and 30 chances of having x, which, by Proposition III, is worth

$$\frac{6a + 30x}{36}$$

to me. This expression being equal to y, and y, after what has gone before, being equal to $36x/31$, it is necessary that

$$\frac{30x + 6a}{36} = \frac{36x}{31} \quad \text{or} \quad x = \frac{31a}{61}.$$

the value of my chance. Consequently the chance of my opponent is worth $30a/61$. The ratio of our chances is therefore 31 : 30.*

The argument is tortuous but he gets there in the end. It was some years—probably with James Bernoulli—before the principles of compound and conditional probabilities were introduced.

Huygens' treatise ends with the words " I finish by stating some propositions " by which he means that he sets out some exercises and gives the answers in an appendix. James Bernoulli set out the solutions in the *Ars Conjectandi*. These exercises are more interesting than the formal propositions because they are possibly the first time that sampling with and without replacement occurs in the literature. Exercise I is just the problem of points all over again, but Exercise II runs as follows :

---

* Presumably opponent's chance $= \dfrac{5}{36} = \dfrac{30}{6.36}$.

His chance is $\dfrac{31}{36} \cdot \dfrac{6}{36} = \dfrac{31}{6.36}$.

Three gamblers A, B and C take 12 balls of which 4 are white and 8 black. They play with the rules that the drawer is blindfold, A is to draw first, then B and then C, the winner to be the one who first draws a white ball. What is the ratio of their chances?

Huygens, as can be seen from his answer, imagines that each ball is replaced after drawing. James Bernoulli pointed out that there were three possible interpretations of the exercise and solved all three. These were (i) the ball can be replaced after drawing, (ii) if it is not replaced after drawing then it may be supposed that the gamblers start with a central pool of 12 balls, or (iii) that they each have 12 balls. A mathematician called John Hudde wrote to Huygens in 1665 pointing out the second interpretation. Montmort, de Moivre and Struyck each discuss the different models. It is not known from where Huygens derived this exercise, and he may have made it up for himself. The third exercise is one of Fermat's problems :

A wagers B that, given 40 cards of which 10 are of one colour, 10 of another, 10 of another and 10 of another, he will draw 4 so as to have one of each colour.*

Exercise IV is again capable of two interpretations (Huygens samples with replacement), while Exercise V is again the problem of points. There is, however, in this exercise possibly the first hint of the gambler's ruin problem, since A and B start with 12 balls and continue to throw three dice on the condition that if 11 is thrown A gives a ball to B and if 14 is thrown B gives one to A. The game is to continue until one or other has all the balls.

Huygens, after the publication of *Games of Chance*, turned to the making of telescopes. He also appears to have travelled extensively in France and England at this time,† and his value as a fertilising agent, spreading the scientific talk of London in Paris and conversely, cannot be overestimated. He was admitted a Fellow of the Royal Society of London in June 1663 at the age

---

* Huygens gives this chance as 1000 to 8139 and so does Todhunter, but I make it 1000 to 9139.

† He was sent by his father to attend the coronation of Charles II, but there was a transit of Mercury on that day and so he never reached the Abbey.

of 34. On his return to Holland he engaged in a correspondence with John Hudde about his Exercises II and IV. Both were right in principle, Huygens sampling with replacement and Hudde without replacement, although Hudde may have made some numerical slips. In 1666 he went to Paris at the invitation of Colbert, Minister of Finance under Louis XIV, to take part in the founding of the Académie des Sciences, and his stay there, with two major interruptions, lasted for fifteen years. He became one of a number of scientists receiving an allowance from Louis XIV and lodging in the Bibliothèque du Roi. Before he went to Paris he received from London in 1662, the year of its publication, a copy of Graunt's *Observations on Bills of Mortality*, in which, among other things, Graunt had tried to construct what may have been the first English life-table. Huygens did not evince any immediate reaction to this at the time, but in 1669 his brother Ludovick wrote to him about it. Ludovick said he had used the data to make a table of life expectations. Christian replied that what he had done could only be approximate, and in a further letter explained the difference between mean length of life and probable length. It was this correspondence which led Christian to write the note *On Examining the calculations of my brother Ludovick* in which he discusses joint lives.

> A man of 56 years of age marries a girl of 16. How long may they expect to live jointly before one or the other dies?

and similar problems.

Christian returned briefly to problems of dice-playing in 1676, the questions being posed by his friend Dierkens concerning the game of Quinquenove. There are 2 dice and 2 players. *A* starts. If he throws a seven he wins, but hands the dice to *B* if he throws anything else. *B* wins if he throws a five or a nine but hands the dice back if he throws anything else, and so on. Huygens used logarithms to get approximate answers and also explored the case of 3 and of 4 dice. This was possibly when he had returned to the Netherlands for a time. His stay in Paris was interrupted by two prolonged visits to Holland, one in 1670–1 and one for nearly two years from 1676–8. He left Paris for ever in 1681.

Sometime in the years 1678–81 he made some calculations on the game of Bassette. There are many versions of the rules of this game, and Huygens may have simplified them to carry out his investigation. His objective seems to have been to find out the Banker's chances so as to know how much to pay for the Bank. The Banker held a complete shuffled pack of 52 cards. The field each held 13 cards of the same suit. The field put down on the table such cards as they wanted to out of their hands and piled on them such stakes as they wished. The Banker put his cards out two at a time, one for himself and one for the field. If his card was a nine (say) he took all the stakes off the nines. If the field's card was a nine (say) he paid the amounts staked on nines. If both cards were the same number he took all the stakes put on that number by the field. Huygens' simplified calculations agree with the more general cases set out by Joseph Saveur and later by James Bernoulli. Various (different) cases were treated by Montmort, John Bernoulli (James' brother), Nicholas Bernoulli (James' nephew), de Moivre and Struyck.

Huygens was a Protestant and he left Paris for the last time in 1681 soon after the revocation of the Edict of Nantes. It is said that efforts were made to persuade him to stay, but he would not and he broke off all contact with French scientists. He still continued in correspondence with Fellows of the Royal Society in London and made a last journey to London in 1689. In 1688 he seems to have occupied himself a little with further gaming problems but produced nothing noteworthy. Somewhere about this time he possibly went a little mad, suffering from an acute persecution mania with regard to his family. At the beginning of 1695 we are told that he lost all his faculties and shortly afterwards he died. He was a celibate all his life and although he did not lead a withdrawn life—in Paris he seems to have had many friends and was not averse to being sociable—he preferred to spend the major part of time either in his study or in his laboratory.

References

Huygens' works and letters have been collected together into twenty volumes (printed in French) and published by the Netherlands Society for the Sciences in 1920. Each volume is well documented and each fresh section is prefaced by explanatory material which gives the biographer nearly all that might be desired. I have supplemented this material where necessary, by the relevant information given by Hoeffer and from reading in his list of references. Huygens, like Galileo, has attracted many biographers, most of whom have been interested in his work as an astronomer and natural scientist.

chapter 12

Wallis, Newton and Pepys

> I do not know what I may appear to the world, but to myself
> I seem to have been only a boy playing on the sea-shore,
> and diverting myself in now and then finding a smoother
> pebble or a prettier shell than ordinary, whilst the great
> ocean of truth lay all undiscovered before me.
>
> Attributed to Sir Isaac Newton by
> D. BREWSTER, *Memoirs*, ii, 27.

John Wallis, man of Kent, was born in 1616 and died in 1703.
He became Savilian Professor of Geometry at Oxford in 1649. An
outstanding mathematician of his day, his book *Arithmetica
Infinitorum* of 1656 did much to clear the way for the invention of
the calculus of fluxions by Newton. From the point of view of
combinatory theory, however, it is his *Algebra* of 1685 which is
the most important of his works. Wallis was chiefly concerned in
this Algebra to take up the cudgels on behalf of his fellow English
mathematicians whose work he thought had been neglected, so
that the book has been described as a contribution to history
rather than to mathematics. The book was also marred to a
certain extent by the violent antipathy which he still felt towards
Descartes, by then thirty-five years in the grave to which Christina
of Sweden had sent him. *A discussion of combinations, alternations
and aliquot parts* is contained in both the English and the Latin
editions of the *Algebra*, and it shows what was the state of the
theory at that time in England. Wallis was concerned first of all
to give an exposition of Buckley's work on combinations. The
chapter is headed " Of the variety of elections or sumptions, in
taking or leaving one or more out of a certain number of things
proposed", and he gives the following exposition :

Let as many numbers as you please be proposed to be combined : suppose five which we will call *a, b, c, d, e.*

1	a	ab	abc	abcd	abcde
2	b	ac	abd	abce	1
4	c	ad	abe	abde	
8	d	ae	acd	acde	
16	e	bc	ace	bcde	
31	5	bd	ade	5	
−5		be	bcd		
26		cd	bce		
		ce	bde		
		de	cde		
		10	10		

Put in so many lines the numbers in duple proportion beginning
with 1. The Sum (31) is the number of Sumptions, or Elections,
wherein one or more of them may several ways be taken. Hence
subduct (5) the Number of the numbers proposed. Because each
of them may be taken once singly. And the remainder (26) shows
how many ways they may be taken in Combination (namely Two
or more at once). And consequently how many products may be
had by the Multiplication of any two or more of them so taken.

It may be thought possibly that this is a liberal interpretation of
Buckley's proposition (see Appendix 1). Wallis also gave the
Arithmetic Triangle and demonstrated how it can be used to
find the number of combinations of *n* things taken *r* at a time. His
chapter immediately following is on permutations, but he added
nothing new here and the two remaining chapters in this section
are on number theory, including the solutions to problems which
Fermat had sent him many years before. Any significant development in combinatory theory thus waits for the advent of James
Bernoulli's researches.

The second half of the seventeenth century is noteworthy not
only to English mathematicians but to all the world as the time
of the flowering of Newton. He was a boy of twelve when Pascal
and Fermat were corresponding, but he knew of Huygens' writings
and it would have been strange had he not known of van
Schooten's *Exercises*. At a time when it was still possible for an
able mathematician to take all knowledge for his province, and

moreover when dicing, and gambling with annuities, were
practised as assiduously in England as anywhere else, it is indeed
strange that not only Newton but nearly the whole of the English
school showed no interest in them.* Nevertheless an examination
of the mathematical works of the period shows that not one of
them gave these matters more than a passing thought. It *must*
be more than accidental that during the whole of the development of probability theory no contribution of any significance was
made by the English.   Such interest as there was expressed itself
in actuarial calculations which, one supposes, could be seen to be
of immediate practical utility ;   no profit was to be found in
discussing the fall of the dice.   This utilitarian view of mathematics was to persist in England through the years.

Mr. Isaac Newton's small contribution to the theory of
annuities appears to have been his sole research on the subject.
His thoughts on dice-playing are derived either from Huygens or
possibly from letters from friends who were once members of
the Mersenne Academy, or both.   These thoughts are set out in
letters which passed between him and Samuel Pepys in 1693 at a
time when James Bernoulli was researching and teaching at a
much higher level.   The letters show that Newton's ideas, clearly
and correctly expressed, were still elementary.   Pepys in London
wrote to Newton in Cambridge on November 22nd, 1693, as
follows :

> The Question :
> *A* has 6 dyes in a box with which he is to fling a 6 ;
> *B* has in another box 12 dyes with which he is to fling 2 sixes ;
> *C* has in another box 18 dyes with which he is to fling 3 sixes.
> Q.—whether *B* and *C* have not as easy a chance as *A* at even luck?

This is yet again the question which caused the Chevalier de
Méré to say that Arithmetic puzzled him because its propositions
were not constant, but he gambled enough to recognise that the
algebraic solution he had reached was different from what he had
actually observed.   Pepys and his friends were also gamblers, but

---

* I except de Moivre, who had his early training under Ozanam in Paris and
who arrived in London already moulded to the pattern he must follow.

possibly not so assiduous as the Chevalier de Méré since, as the letters show, they had no idea of the answer. This Question was sent to Newton by the hand of Mr. Smith, the writing-master at Christ's Hospital, and in the letter which accompanied it Pepys explained that it was Mr. Smith who wanted to know the answer.

due to his soe earnest an application after truth.

There is a hint of the need to justify interest in these problems to Newton. This may be, of course, the reason why Newton did not research in the probability calculus ; he may have disapproved of gambling, but Pepys thought he would not object to a search for truth.

That it was Pepys and not Mr. Smith who really wanted to know the answer is apparent almost immediately. Pepys' ingenuousness probably did not deceive Newton since he found, by cross-examining Mr. Smith in order to clarify the Question, that he didn't know much about it. At any rate Newton set himself to work and on November 26th (1693) wrote back at length to Pepys.

In reading the Question it seemed to me at first to be ill stated. . . .

He points out the difference between being required to throw just one six or at least one six and concludes that it must be the latter.

If this problem must be understood according to the plainest sense of the words, I think the sense must be this : . . . .

What is the expectation or hope of A to throw every time one six at least within six dyes?

What is the expectation or hope of B to throw every time two sixes at least with twelve dyes?

What is the expectation or hope of C to throw every time three sixes or more than three with 18 dyes?

And whether have not B and C as great an expectation or hope to hit every time what they throw for as A hath to hit what he throws for?

It is possibly a little strange that he should have preferred the " at least " interpretation instead of the " just one." Mr. Smith may have helped his interpretation, although from Newton's

WALLIS, NEWTON AND PEPYS

remarks it might seem that Mr. Smith did not altogether agree with it. Newton continues his letter by saying that if the construction he has put on the question is correct, then he has done the computations and will send them if Pepys wants them. Pepys wrote again on December 9th. He said that Newton's interpretation of the Question was the right one, and that Mr. Smith's interpretation was not " in any wise warrantable from the terms of it " and he asks for the computations. He also restates the question :

> I should avoid some of the ambiguities that commonly hang about our discoursings of it, by changing the characters of the dice from numbers to letters and supposing them instead of 1, 2, 3, . . . etc. to bee branded with the 6 initiall letters of the alphabet A, B, C, D, E, F. And the case should then bee this :—
> Peter, a criminal convict doomed to dye, Paul his friend prevails for his having the benefit of one throw only for his life, upon dice soe prepared ; with a choice of any one of these three chances for it, viz.:
> One F at least on six such dice,
> Two F's at least on twelve such dice,
> or
> Three F's at least on eighteen such dice.
> Question : Which one of these chances should Peter in this case choose ?

Some if not all of the French mathematicians who were interested in gaming problems could have solved this problem in 1654. Certainly Huygens would have been able to demonstrate the answer and Newton does not go beyond this. He does not attempt anything explicit in the way of binomial coefficients but proceeds to enumerate the possibilities by a method obviously borrowed from Wallis. He sent the following to Pepys on December 16th, 1693 :

> I set down the following progressions of numbers :—

Progr. 1	1	2	3	4	5	6	the number of the dice.
Progr. 2	0	1	3	6	10	15	

Progr. 3	6	36	216	1296	7776	46656	the number of all the chances upon them.
Progr. 4	5	25	125	625	3125	15625	the number of chances without sixes.
Progr. 5	1	5	25	125	625	3125	
Progr. 6	1	10	75	500	3125	18750	chances for one six and no more.
Progr. 7		1	5	25	125	625	
Progr. 8		1	15	150	1250	9375	chances for two sixes and no more.

The progressions in this table are thus found : the first progression which expresses the number of the dice is an arithmetical one. . . . The second is found by adding to every term the term of the progression above it . . . and by these rules the progressions may be continued to as many dice as you please. . . . Since A plays with six dice . . . I consult the numbers in the column under 6, and there from 46656 . . . I subduct 15625 . . . and the remainder 31031 is the number of all chances with one six or above.

For 12 dice (calculations not shown) he gets the number 13467,04211/21767,82336 (which is correct) and stops there. The arithmetic would have been formidable for 18 dice and Huygens would probably have used logarithms. Newton finishes by remarking that the chances $A : B : C$ are in descending order of magnitude.

Pepys possibly did not like this answer because he wrote yet again on December 21st, 1693 to ask whether it mattered if he threw all the 12 dice at once or threw two dice-boxes each with 6 dice in them or had two throws with one dice-box containing 6 dice. Newton replied immediately on December 23rd. He wrote at length and with clarity but was, it would seem, tired of the problem. The letter begins :

> Sir, I take it to be the same case whether a man, to throw two sixes, have one throw with twelve dyes or two throws with six, but I reccon it an easier task to throw with six dyes one six at one throw than two sixes at two throws . . .

In addition to asking Newton for his views on the Question, Pepys appears also to have consulted his friend Mr. George Tollet who has been described as a Commissioner for the Navy and Secretary to the Commissioner for Excise. Mr. Tollet demolished the argument that there was any advantage to be gained from throwing with two dice-boxes, and from his calculations seems to have a workmanlike knowledge of how to use the binomial theorem combinatorially.* Pepys' nephew, John Jackson, also took a hand. The episode closes with a letter from Pepys to Tollet on February 14th, 1694.

> Sir : I have taken all this time for 't, and can scarce yet tell in what terms to thank you for your late admirable present ; whether as a further exercise on my old lesson . . . or for a new and different one, namely that *A* has an easier task than *B* and a yet more easy one than *C*, such I take it being the doctrine of this paper ; and full glad I am of my soe seasonably meeting with it, as being upon the very brink of a wager, 10 L. deep, upon my former belief. But apostacy wee all know is now no novelty, and therefore like others I shall endeavour to make the best of mine, and face my antagonist down that I always meant this . . . .

The fiction of Mr. Smith's thirst after truth is at last explained !

References

The letters from which I have worked are—
Private Correspondence and Miscellaneous Papers of Samuel Pepys (J. R. Tanner, Editor).
These appear to be exhaustive. The Newton–Pepys letters are also published in the Newton letters at present being issued by the Royal Society of London.

The substance of this chapter was given by me in an article in the *Annals of Science, 1958* (" Mr. Newton, Mr. Pepys and Dyse ").

* It is possible that Newton would also have proceeded much as Tollet did if he had thought that Pepys would have understood him.

chapter 13

James Bernoulli and "Ars Conjectandi"

The years like great black oxen tread the world,
And God the herdsman goads them on behind,
And I am broken by their passing feet.
                    w. b. yeats, " The Countess Cathleen ", sc. v.

The Bernoullis came to Basel, after a brief sojourn in Frankfurt-am-Main, sometime in the early sixteen hundreds, refugees from the city of Antwerp where those of the Protestant faith did not find the government of the Duke of Alba easy or even comfortable. They established themselves as merchant bankers in the town, and Nicholas (1623–1708) was for a time head of the civic government. James, the eldest son of Nicholas, was born in Basel on Christmas Day, 1654, the year of the Fermat–Pascal correspondence, but before the visit of the young Huygens to Paris. He was the first of the learned group of scholars of that name, now commemorated by the University of Basel in the building known as the Bernoullianum. James was destined by his parents to be a minister of the Reformed Church, and took his degree in theology, for which he had also to study the humanities, at Basel. At some period about this time it is said that he found a copy of Euclid and on studying it became desirous of being a mathematician.* What first set him on the mathematical path may be a matter for conjecture, but there is no doubt that at the age of 18 he was closely interested in astronomy.† By the time he was twenty-two (1676) and had set off on the first of his

---

* One may note also the story about Pascal.

† Later in life he took as his crest Phaeton in his sun chariot with the motto *Invito patre, sidere verso* (Against my father's will I study the stars).

130

journeys, he was obviously moulded to his life-pattern of teacher and scholar. At Geneva he instructed a blind girl in mathematics, writing a paper with the title " On Teaching Mathematics to the Blind," and at Bordeaux he drew up some tables connected with sun-clocks. He arrived back in Basel in 1680 in time to observe the comet of that year. This set him further on the path of astronomy ; he studied Descartes' writings intensively and after some prolonged calculations predicted the date of the comet's return. This being accomplished he almost immediately set out again on his travels, going north this time to the Low Countries and to England. He made scientific acquaintances wherever he went and he learnt assiduously, staying in any one place until he had mastered what it had to offer.

On returning to Basel in 1682 at the age of 27 he seems to have finished with travel. All his life he was a teacher as well as being eminent in mathematical research, and he now started his professional career as a teacher in a school, presumably distilling the fruits of his new learning. He corresponded with Leibnitz and appears to have been one of the first to learn and apply to problems the new differential calculus. In 1684 he married, and in 1687, on the death of Peter Megerlin, he was appointed to the chair of mathematics at the University of Basel, holding this chair until his death on August 16th, 1705. It is said that he was so successful as a teacher that students from all over Europe came to study under him. Many scientific and academic honours were bestowed on him, but in spite of the recognition by his contemporaries of his achievements as a mathematician it is to be doubted whether his years in the chair of mathematics were happy ones. John Bernoulli, the brother of James, was also interested in mathematics and was extremely jealous of his brother. Even in the disputatious and quarrelsome seventeenth and eighteenth centuries the Bernoulli brothers stand out by the bitterness and vehemence of their controversies. The method of challenge to the problem, which seems to have faded in the French academies into a means of disseminating information, was revived and developed by James and John as a method of intellectual warfare, with the two brothers usually on opposing sides. Leibnitz, instead of being a peacemaker, seems to have fostered and encouraged this strife. From it, however, came a number of the mathematical developments, including the origin and delineation of the calculus of variations, most of these, possibly, being ascribable to James.

At some time early in his career as a mathematician James became interested in the calculus of probability. It is not known when, and Leibnitz's statement that it was at his insistence that James began to study the problems involved can probably be discounted. James was profoundly influenced by Huygens' book *De Ratiociniis in Aleae Ludo* and may have come to this while working through van Schooten's *Exercises* as a student. There is no record of his having met Huygens on his visit to the Low Countries, but this is not an impossibility. Huygens at that time had returned to The Hague from Paris, and he and James would have had a common interest in astronomy. All that is definite is that sometime before 1685 he must have started work on the problems connected with games of chance, for in that year he published a number of papers on this subject in the *Journal des Scavans*. Most of the problems in these papers were obviously inspired by Huygens' work.* These papers are not commonly cited, but Scott (*History of Mathematics, q.v.*) gives a typical example.

A and B play with a die, the one that throws an ace first being the winner. A throws once, and then B throws once also. A then throws twice and then B. A then throws three times and then B and so on. What is A's expectation of winning?

James did not give the solution of this particular problem until his paper in the *Acta Eruditorum* of 1690.

---

* That his interest in games of chance was an early one is borne out by his notebooks. James Bernoulli was a methodical man and from the beginning kept a record of all the problems on which he worked and of their solutions. All but the probability problems are in Latin, but these latter are in French. It is unfortunate perhaps that the University of Basel does not allow free access to these notebooks. A study of what James was working on at any given time would prove of great interest. For, because of the students who came to study with him, we might gain some idea of when his ideas about probability began to percolate throughout Europe.

It appears likely that the work for which he is best known among probabilists, the *Ars Conjectandi*, was written during this period of his life, and that he held back publication while he pondered over the implications of the theorem which now goes by his name. He had considerable correspondence with Leibnitz over it, possibly the first disputation on inverse probability which is recorded. James stuck stubbornly to his point of view, but may have had inward doubts because he did not publish. When he died in 1705 the manuscript was all but complete, and Nicholas, the son of James' brother Nicholas, was asked to edit it with a view to publication. Nicholas was at that time only 18. He had been a pupil of James and was probably the best-fitted person to do the editing, but it was a great deal to expect from a young man and he may have been afraid of attack from Leibnitz. As the years pass, however, his reticence becomes a little puzzling. He wrote on the part probability could play in jurisprudence for his doctoral thesis in 1709, this stemming from James' ideas about the probability of propositions. During the years 1712–13 he carried on a long and interesting correspondence on probability with Montmort, showing that he was a probabilist of some stature. It would seem possibly that the animosities between James and John and the envy which John did not cease to feel even after his brother's death, were natural to the whole family. However by 1713 it was not possible to delay further. The mathematical world had been waiting for *Ars Conjectandi*, at least since Fontenelle gave a summary of the contents in his *Éloge* of James, and the pressure of public opinion must have been growing. Nicholas wrote a preface for James' book explaining the contents and the reason for the length of his delay, and eight years after the death of James it appeared. The preface runs as follows :

> At last, here is the art of conjecture, the posthumous treatise of my uncle which has been so long awaited. The Brothers Thurnisius, thinking to do a public service, have acquired the manuscript from the author's executors, and have printed it at their own expense. The author wanted to make known in civil life the usefulness of that part of mathematics which is directed towards the measurement of probabilities. We have already seen in the memoirs of the

Academy of Sciences of 1705* and in the Scientific Abstracts of
Paris of the year 1706, by what method and up to what point the
author has fulfilled the task which he set himself. He has divided
his work into four parts. The first contains the treatise of the
illustrious Huygens, " Reasoning on Games of Chance," with
notes, in which one finds the first elements of the art of conjecture.
The second part is comprised of the theory of permutations and
combinations, theory so necessary for the calculation of probabilities and the use of which he explains in the third part for
solution of games of chance. In the fourth part he undertook
to apply the principles previously developed to civil, moral and
economic affairs. But held back for a long time by ill-health, and at
last prevented by death itself, he was obliged to leave it
imperfect. The editors would have liked the brother of the author
(i.e. John), so capable of achieving this work, to have taken over
completing it, but they knew that he had undertaken so much that
they did not even ask him about it. As they knew that in an
inaugural dissertation I had given some trial to this theory as
applied to law, they asked me to undertake the completion. But
my absence on travels did not allow me to do this. On my return
they asked me again and I think I ought to mention why I did not.
I was too young and inexperienced to know how to complete it.
I did not feel I had enough initiative and I was afraid not only
that I would not hold the attention of the reader but even that by
risking the possibility of adding trivial and ordinary things I would
do wrong to the rest of the work. The printing of this treatise being
already fairly well advanced, I advised the printers to give it to
the public as the author left it. However, as it is necessary that so
useful a thing as the application of probabilities to economic and
political affairs should not be forgotten, we beg the illustrious
author of " Essay on the Analysis of Games of Chance " (i.e.
Montmort) and the celebrated de Moivre who wrote a little
time ago some excellent fragments of this art, to set themselves to
this work and to consecrate to it a little of the time that they set
aside for the public good. We hope especially that the generalisations given by the author in the five chapters of the last part
will offer to the reader the principles of application important for
the solution of particular problems. This is all I have to say on

* Fontenelle's *Éloge*.

this treatise. The editors have added to it the theorems on infinite series which the author made the subject of five dissertations and which are out of print. It was for this reason that they have reprinted them at the end of this work. The affinity of the subject-matter has made us also add the paper written in French by the author entitled "Letter to a friend on chances in the game of tennis".

The comments and additions by James to Huygens' treatise are of value. It has been noted that in his treatise Huygens set out what must have been the first mention in print of the gambler's ruin problem. James generalised it to the case where A has m units to stake, B has n and the chances of $A : B$ for winning a game are $a : b$. He showed that A's chance of winning a set of games by encompassing B's ruin is

$$\frac{a^n(a^m - b^m)}{a^{m+n} - b^{m+n}}$$

and B's similar chance is

$$\frac{b^m(a^n - b^n)}{a^{m+n} - b^{m+n}}$$

There are several generalisations of this kind which amply demonstrate his powerful mathematical attack on such problems. In the section on permutations and combinations he gives us the arithmetic triangle, the formula for the number of combinations of n things taken r at a time, the expansion of the binomial series for a positive integral index, and a method of summing the rth powers of the first n natural numbers for $r = 1, 2, \ldots, 10$ by means of an expression using the now famous Bernoulli numbers. It seems likely that James had some idea of using at least part of the *Ars Conjectandi* as a textbook for his students since he is concerned not only to give correct solutions but on occasion to point out errors in reasoning which may lead to wrong answers. An example of this is found in a demonstration of the simple trap of averaging over expectations which do not have the same weight, and it is interesting to note James' powers and insight as a teacher in that he troubled to point this out.

The first three parts of the *Ars Conjectandi* alone would have been enough to establish James' reputation as a probabilist since he was an incomparably better algebraist than Huygens and had a much more profound insight. This may have been because he was a quarter of a century later, and algebra was developing with some speed at this time ; but I would think not. There is a clarity and a purposiveness about James' attack on any problem which Huygens never had. It was, however, the fourth part of his book *Pars Quarta: tradens Usum et Applicationem Doctrinae in Civilibus, Moralibus et Oeconomicis* which was to prove Pandora's box. In this part is stated what James described as his " golden theorem", and about which he wrote :

> This is therefore the problem that I now want to publish here, having considered it closely for a period of twenty years, and it is a problem of which the novelty, as well as the high utility, together with its grave difficulty, exceed in weight and value all the remaining chapters of my doctrine.

As a prelude he considers the following proposition. Let r, s, n and t all be positive integers with $t = r + s$. Consider the expansion of $(r + s)^{nt}$. Let the sum of the $(2n + 1)$ terms, made up of the largest term and the n immediately before it and the n immediately after it, be u, and consider the ratio $u : (r + s)^{nt} - u$. For any free choice of n this ratio may be made as large as desired. If it is desired that the ratio shall not be less than c then n must be equal to

$$\left(1 + \frac{s}{r + 1}\right) \left(\frac{\log c + \log (s - 1)}{\log (r + 1) - \log r}\right) - \frac{s}{r + 1}$$

or

$$\left(1 + \frac{s}{r + 1}\right) \left(\frac{\log c + \log (r - 1)}{\log (s + 1) - \log s}\right) - \frac{r}{s + 1},$$

whichever is the greater. James took a long time to establish this because he had not thought of how to expand a factorial, but his proof is correct. If we write

$$(r + s)^{nt} = t^{nt} \left(\frac{r}{t} + \frac{s}{t}\right)^{nt}$$

then r/t can be looked on as the probability that an event will happen in a single trial and s/t as its complement. The sum of the middle $2n + 1$ terms will be proportional to the probability that in nt trials the number of " happenings " of the event will be between $nr - n$ and $nr + n$. The ratio therefore of the limits for the event to the total number of trials will be $(r - 1)/t$ and $(r + 1)/t$. n is now chosen to give the required odds of $c : 1$ and the result follows.

So far this is a straightforward argument in direct probability, and to begin with James treats it in this way. He gave a numerical example in which he took $r = 30$, $s = 20$ ($r + s = t = 50$), with c to be 1000. For a probability of $999/1000$ that the ratio of the number of times that the event happens to the total number of events should lie between $31/50$ and $29/50$ it is enough to make 25,550 (actually 25,500) trials. For the chance to be $9999/10,000$ it is enough to make 31,258 trials and so on. He then proceeds to turn the argument round. First, he says (quite correctly), suppose an urn with black and white balls in the proportions 3 white to 2 black. Let the sampling be with replacement. If 25,500 trials are made of the drawing of a single ball with the ball replaced after each trial the chance is only $1/10^3$ that the proportion of white balls observed lies outside the limits $29/50$ and $31/50$. Now suppose that nothing is known about the composition of colours within the urn. A large number of drawings is made with replacement as before, with the result that a white ball is seen R times and a black ball S times. The proportion of white balls is then inferred to be $R/R + S$. James justified himself with an appeal to Fate.

> If thus all events through all eternity could be repeated, by which we would go from probability to certainty, one would find that everything in the world happens from definite causes and according to definite rules, and that we would be forced to assume amongst the most apparently fortuitous things a certain necessity, or, so to say, FATE. I do not know whether Plato in his theory about the circulation of things wanted to hint at this when he affirms that in the course of unnumbered centuries everything reverts to its original state . . . .

James has definitely started here the controversy on inverse probability, a controversy which, looking back, appeared inevitable sooner or later. He was responsible for many of our " modern " ideas and, nearer to his own time, he can be accounted responsible for de Moivre's derivation of the normal curve limit to the sum of a number of binomial probabilities. For de Moivre found this limit in order to refine the " golden theorem ". Two quotations will serve to show how the yeast worked. They are both taken from de Moivre's *Doctrine of Chances*.

> And thus in all cases it will be found, that although Chance produces Irregularities, still the odds will be infinitely great that in the process of time, those Irregularities will bear no proportion to the recurrency of that Order which naturally results from ORIGINAL DESIGN.

And again in defence of James :

> I shall only add that this method of Reasoning may be usefully applied in some other very interesting enquiries and shall conclude this remark with a passage from the *Ars Conjectandi* of Mr. James Bernoulli, where that acute and judicious writer thus introduceth his solution of the Problem for *Assigning the Limits within which, by the repetition of experiments, the Probability of an Event may approach indefinitely to a given probability*. This, says he, is the problem I now impart to the Public, after having kept it by me for 20 years : new it is and difficult, but of such excellent use that it gives a high value and dignity to every other branch of this doctrine. . . . Yet there are writers, of a class indeed very different from that of James Bernoulli, who insinuate as if the Doctrine of Probabilities could have no place in any serious enquiry and that studies of this kind, trivial and easy as they be, rather disqualify a man for reasoning on every other subject. Let the Reader choose.

James was buried in the Baarfüsser Kirche in Basel. This is now a museum, but his tombstone has been removed and may be seen in the cloisters of the Münster, on top of the hill. What James himself considered to be his principal discovery was engraved on his tombstone. In his paper on the logarithmic spiral he wrote :

> Because our wonderful curve always in its changes remains

constantly the same and identical in type it can be regarded as the symbol of fortitude and constancy in adversity : or even of the resurrection of our flesh after various changes and at length after death itself. Indeed if it were the habit to imitate Archimedes today I would order this spiral to be inscribed on my tomb with the epitaph *Eadem mutata resurgo* (However changed it is always the same).*

This was done.

References

A great deal of material concerning the Bernoulli family has been published in Basel (in German). Hoeffer gives a summary of what was known up to the year in which the *Biographie Universelle* was written (1854). The *Éloge* of James Bernoulli by Fontenelle is useful both for background material and for mathematical information. I would also draw attention to

J. F. SCOTT, *History of Mathematics*

and an article by K. PEARSON on "James Bernoulli's Theorem" in the journal *Biometrika*. The Bernoulli letters are to be printed in full and from the contents of the first volume should prove helpful in further understanding of this able family.

---

* Translation given by K. Pearson in a lecture on James Bernoulli's Theorem. Other translations in this chapter were made with the aid of Jean Edmiston.

chapter **14**

Pierre Rémond de Montmort and the "Essai d'Analyse"

Patience, and shuffle the cards.

Don Quixote, ii, 23.

The latter part of the seventeenth century is noteworthy in the history of mathematics. Not only was Newton, the greatest all-round mathematician of all time, at his most productive, but all over Europe there were mathematicians, who were his equals in their own field, pouring out their ideas about algebra and the new calculus. And while the influence and example of Newton (1642–1727) made itself felt wherever the natural philosophers were working, possibly an even greater arbiter of destiny was the philosopher Leibnitz (1646–1716). Opinions may differ concerning his creative powers as an analyst, but he appears to have acted as a focus for the European mathematicians, both as a critic and as a disseminator of their suggested new theorems. His influence on the Bernoulli brothers and on the group of mathematicians working in Paris cannot be overestimated. Thus, while he did not start James working on the probability calculus, it was to him that James turned when he created his " golden theorem," and it was (possibly) Leibnitz's criticism of it which kept James and later his nephew Nicholas from its publication (see page 133). He did not contribute to the calculus of probabilities itself, but there is no doubt that his exposition of the differential calculus and his work on the exponential series—to quote only two of his contributions—made the way possible for the further advances in probability theory which came with comparative suddenness.

Applications of new mathematical techniques to problems of

observational data always show a time-lag, and in no subject is this evinced more clearly than in the probability calculus. James Bernoulli, teaching and researching at Basel between 1685 and 1705, distilling the mathematical ideas of Leibnitz and adding many of his own, was the first to appreciate how the then " modern " analysis could be used in the analysis of games of chance, and without him it is doubtful whether Pierre-Rémond de Montmort (1678–1719) or Abraham de Moivre (1667–1754) would have contributed as much as they did. For Montmort took up James' ideas from his papers of 1685 onwards and carried out a painstaking analysis of various games of chance using the mathematical analysis known to him, while de Moivre, whose first ideas on the subject were to a certain extent reflections of Montmort and Bernoulli, finally achieved his full stature by generalising and extending their work. Montmort is an attractive character to whom, in the writer's opinion, insufficient attention has been paid. For he undoubtedly inspired de Moivre, and from his correspondence with Nicholas Bernoulli we gain a real appreciation of James, who was Nicholas' teacher.

The life of Pierre-Rémond de Montmort, after a stormy start, was a simple, happy one. He was born in Paris on October 27th, 1678, the second of three sons of François and Marguerite Rallu Rémond who were of the nobility. François Rémond intended his son to study law, there being a vacant magistracy waiting for him, but, his biographer says, the son was contemptuous of any restraint, and left home, travelling first to England and then to the Low Countries and Germany. At the house of his cousin in Germany he found a copy of Malebranche's *La Recherche de la Vérité*, and on reading it appears to have suffered a minor conversion. He was inspired to return home and make his peace with his father. This was in 1699. His father died shortly afterwards leaving him, at the age of 22, a large fortune. He does not, however, appear to have plunged into the dissipated life which was thought natural for a young nobleman of this time to lead. His conversion stood him in good stead and he continued to occupy his time with " pious exercises " and with studying philosophy and mathematics with Father Nicholas de Malebranche who was then in the House of the Oratory of Saint-Honoré de Paris. After learning some mathematics Montmort went to England again (in 1700), on purpose to make the acquaintance of the English mathematicians and in particular of Newton. This visit gave further impetus to his study of mathematics, and he came back to Paris to pursue his studies in algebra, geometry and the new calculus, which he found " thorny ".*

Montmort succeeded his elder brother as a canon of Notre-Dame about this time, but he did not occupy his stall for long. He bought the estate of Montmort in 1704 and two years later (1706) he resigned his canonry in order to marry the great-niece of the Duchess of Angoulême. His mathematical interests had suffered to some extent from his ecclesiastical duties, but during the time of his canonry he had printed at his own expense several scientific works, including Newton on the quadrature of curves. After his marriage he settled down on his country estate and set himself to work on the theory of probability. Quite why he chose this topic is not known, for he was no gambler. It was known in France and known to Montmort, if only from the éloge of James Bernoulli, that James had, at the time of his death, left the manuscript of a book on the subject. Possibly Montmort had a contact with a pupil of James—it is thought he did not meet Nicholas until 1709—and this contact inspired him to pursue the new calculus with its fascinating sidelines of the summation of infinite series and the manipulation of binomial coefficients. It seems unlikely that he would have taken it up without some impetus of this kind. He himself says :

> Several of my friends have urged me for a long time to see if I could not determine by algebra what is the advantage of the Banker in the game of Pharaoh. I would not have dared undertake this research . . . if the success of M. Bernoulli had not incited me for more than two years to try to calculate the different chances in this game. . . . This gave me the idea of getting to the bottom of this matter and the desire to make up to the Public in some

---

* One may speculate here about the new calculus. Montmort may have heard talk of it in London, but he uses Leibnitz's *d* notation which would imply that he learnt it from a non-English source.

fashion for the loss that it has sustained in being deprived of the excellent work of M. Bernoulli. . . . *

The results of his researches were published in the *Essai d'Analyse sur les Jeux de Hasard* printed in Paris in 1708. This, the first edition, often passed over because of the greater length of the second edition and the incorporation of the Montmort–Bernoulli correspondence, is worthy of comment. As I have noted, one does not know quite what started Montmort off, but it is known that he wrote it in the comparative isolation of his country estate and it represents therefore the attempts of a competent mathematician, given the outline of the problems discussed by James Bernoulli, to solve them using the latest mathematical techniques, and to extend the field of application of these techniques to card games. This first edition is Montmort himself speaking ; the second edition, while much more comprehensive, is probably a mixture of Montmort and Nicholas Bernoulli. The fact that he wrote at all is probably a fortunate one for the probability calculus. That a nobleman of France and an ex-canon of Notre-Dame should find such problems worthy of speculation and not an impious study would give the subject a certain cachet and air of respectability which left lesser mortals free to work undisturbed. Even so, Montmort found it necessary to write an apologia for having spent his time working on such problems and he wrote this with a self-conscious rectitude :

> It is particularly in games of chance that the weakness of the human mind appears and its leaning towards superstition . . . .There are those who will play only with packs of cards with which they have won, with the thought that good luck is attached to them. Others on the contrary prefer packs with which they have lost, with the idea that having lost a few times with them it is less likely that they will go on losing, as if the past can decide something for the future. . . . Others refuse to shuffle the cards and believe they must infallibly lose if they deviate from their rules. Finally there

---

* Since he had published a number of the scientific works of others at his own expense it is a little curious that he did not try to publish *l'excellent ouvrage de M. Bernoulli*.

are those who look for advantage where there is none, or at least so small as to be negligible. Nearly the same thing can be said of the conduct of men in all situations of life where chance plays a part. It is the same superstitions which govern them, the same imagination which rules their method of procedure and which blinds their fears and hopes. Often they abandon a small, certain, wealth in order to run fearfully after a greater : and often they wilfully give up well-founded expectations in order to conserve a thing the value of which is nothing like those which they neglect. The general principle of these superstitions and errors is that most men attribute the distribution of good and evil and generally all the happenings in this world to a fatal power which works without order or rule. They believe that it is necessary to appease this blind divinity that one calls Fortune, in order to force her to be favourable to them in following the rules which they have imagined. I think therefore it would be useful, not only to gamesters but to all men in general, to know that chance has rules which can be known, and that through not knowing these rules they make faults every day, the results of which with more reason may be imputed to themselves than to the destiny which they accuse. . . . It is certain that men do not work honestly as hard to obtain what they want as they do in the pursuit of Fortune or Destiny. . . . The conduct of men usually makes their good fortune or their bad fortune, and wise men leave as little to chance as possible.

There is much more along these lines.

In this first edition of the *Essai d'Analyse* Montmort begins by finding the chances involved in various games of cards. He discusses such simple games as Pharaoh, Bassette, Lansquenet and Treize, and then, not so fully or successfully, Ombre and Picquet. The work is easy to read in that he prefaces each section with the rules of the game discussed, so that what he is trying to do can be explicitly understood. Possibly he found it necessary to do this because different versions of the games were in vogue, but this does not always occur to other writers. Having set down the rules, he solves simple cases in a method somewhat reminiscent of Huygens, and then takes a plunge into a general solution which appears to be correct but is not always demonstrably so. The *Problèmes divers sur le jeu du treize* are interesting indeed in that

he gives the matching distribution and its exponential limit. Treize
has survived today as the children's game of Snap.

The players draw first of all as to who shall be the Bank. Let us
suppose that this is Pierre, and the number of players whatever one
likes. Pierre having a complete pack of 52 shuffled cards, turns them
up one after the other. Naming and pronouncing one when he
turns the first card, two when he turns the second, three when he
turns the third, and so on until the thirteenth which is the King.
Now if in all this proceeding there is no card of rank agreeing with
the number called, he pays each one of the Players taking part and
yields the Bank to the player on his right.

But if it has happened in the turning of the thirteen cards that
there has been an agreement, for example turning up an ace at the
time he has called one, or a two at the time he has called two, or
three when he has called three, he takes all the stakes and begins
again as before calling one, then two, etc.

It may happen that Pierre, having won several times and
beginning again at one has not enough cards in his hand to complete the thirteen, etc., etc.

He begins by assuming Pierre has two cards and one opponent,
Paul. Then Pierre has three cards, four, and finally any number.
Next he argues generally by building up his chances. Thus,

If we call S the chance we want, the number of cards Pierre
has being denoted by p, and let g be Pierre's chance when the
number of cards he holds is $p - 1$, and let d be the chance when
his number of cards is $p - 2$, we have

$$S = \frac{g(p - 1) + d}{p}$$

He gives p successively values $1, 2, \ldots, 13$ and calculates Pierre's
chance at each stage. It is, however, the remarks on this which are
interesting. After his calculations he says :

The preceding solution furnishes a singular use of the figurate
numbers (of which I shall speak later), for I find in examining the
formula, that Pierre's chance is expressible by an infinite series
of terms which have alternate $+$ and $-$ signs, and such that the
numerator is the series of numbers which are found in the Table

(i.e. the Arithmetic Triangle) in the perpendicular column which corresponds to p, beginning with p, and the denominator the series of products $p \times p - 1 \times p - 2 \times p - 3 \times p - 4 \times p - 5$; so that, cancelling out the common terms, we have for Pierre's chance the very simple series

$$\frac{1}{1} - \frac{1}{1.2} + \frac{1}{1.2.3} - \frac{1}{1.2.3.4} + \frac{1}{1.2.3.4.5} - \frac{1}{1.2.3.4.5.6} + \text{etc.}$$

Let us suppose a logarithm of which the subtangent is unity. We will take on this curve a constant ordinate $= 1$ and another ordinate smaller $= 1 - y$. We will call x the abscissa contained between these two ordinates and we will have $dx = dy/(1 - y)$ and $x = y + \frac{1}{2}y.y + \frac{1}{3}y^3 + \frac{1}{4}y^4 + \text{etc.}$, whence by the method of inversion of series

$$y = x - \frac{x.x}{1.2} + \frac{x^3}{1.2.3} - \frac{x^4}{1.2.3.4} + \frac{x^5}{1.2.3.4.5} - \text{etc.}$$

which, putting $x = 1$, becomes

$$1 - \frac{1}{1.2} + \frac{1}{1.2.3} - \frac{1}{1.2.3.4} + \frac{1}{1.2.3.4.5} - \text{etc.}$$

He gives a more general way of getting this series which he says he has obtained from a paper of Leibnitz (Leipzig, 1693) in which is the problem *Un logarithme étant donnée, trouver le nombre qui lui correspond.* This is possibly the first exponential limit in the calculus of probability, but having achieved it Montmort can't make much use of it. He contents himself by remarking

> One could make several interesting remarks about these series but that would take us outside the present subject and would lead us too far away.

In the second half of the first part on Piquet, Ombre, etc. he interpolates a section on problems in combinations. This is all quite sound mathematics, although he takes a very long time to establish the Arithmetic Triangle. The principle of conditional probability, often attributed to de Moivre but probably dating back to the controversy between Huygens and Hudde, is used with facility and understanding. He illustrates this principle by considering a pack of 40 playing cards, " mêlées à discrétion," the court cards being excluded. If the pack is dealt, the chance that the first four cards will be the four aces is established as

$$\frac{1.2.3.4}{40.39.38.37}.$$

The generalisation of this idea causes some difficulty and certainly calls for some sort of notation. Thus (Proposition XIV),

> Let there be any number of cards whatever composed of an equal number of aces, of twos, of threes, of fours, etc. Pierre wagers that in drawing a given number of cards from this pack at random he will have so many singletons, so many doubles, so many triples, so many quadruples, so many quintuples, etc.

He lets m be the total number of cards, q the number of aces, etc., p the number of different kinds ($q \times p = m$), b is the highest number in a group, i.e. doubles or triples or quadruples, etc. of the cards which Pierre says he wants to draw, $c < b$ the next number, $d < c$ and so on, while B is the number of the groups of cards called b, etc.

> I will express also by the symbol

> the number which expresses in how many ways b may be taken in $q, \ldots$ raising this number by the index B, and multiplying the result by the number which expresses the number of ways B may be taken in p, etc. . . .

The total number of ways of reaching the desired event is given as

$$S = \Box_{b}^{q_{B}} \times \Box_{B}^{p} \times \Box_{c}^{q_{C}} \times \boxed{}_{C}^{p-B} \times \Box_{d}^{q_{D}} \times \boxed{}_{D}^{p-B-c} \times \Box_{e}^{q_{E}} \times \boxed{}_{E}^{p-B-C-D} \times \text{etc.}$$

this to be divided by the total number of ways in which the number of cards to be drawn may be taken from m. His exposition is not entirely clear, but the boxes are just the combinatorial coefficients. This may be seen immediately from his first example and the correctness of his generalisation then realised.

Example I.—If Pierre proposes to draw seven cards from fifty-two, so that he will have three doubles and a single,* the formula gives

$$\frac{\underset{2}{\overset{4_3}{\square}} \times \underset{3}{\overset{13}{\square}} \times \underset{1}{\overset{4_1}{\square}} \times \underset{1}{\overset{10}{\square}}}{4 \times 13 \times 17 \times 10 \times 7 \times 47 \times 46} = \frac{16632}{900473}$$

It is possibly fortunate that this notation was not generally adopted.

In the second part of his treatise Montmort discusses the game of Quinquenove and the game of Hazard, remarking about the latter that the game is known only in England. This, from the references in medieval French literature, is unlikely, but since the rules have been a matter for discussion we set down here Montmort's version of them.

> This game is played with two dice like Quinquenove. Let us call Pierre the die-thrower and suppose Paul represents all the other players. Pierre throws the dice until he has obtained either 5, 6, 7, 8 or 9. Any of these numbers, whichever turns up first, is Paul's chance. Then Pierre throws the dice again to obtain his own chance. He may have the numbers 4, 5, 6, 7, 8, 9 or 10, so that he has two more possibilities than Paul, namely 4 and 10.
> (i) If Pierre, after having given Paul a chance which is 6 or 8, throws with his second the same number, or twelve, he wins. But if he throws two aces, or a two and an ace, or eleven, he loses.
> (ii) If he has thrown for Paul the number 5 or 9, and in the following throw he gets the same number he wins : but if he throws two aces or a two and an ace, or eleven or twelve, he loses.
> (iii) If he has given Paul the chance 7, and he throws the same number at the next throw, or eleven, he wins. But if he throws two aces, a two and an ace, or twelve, he loses.
> (iv) Pierre having obtained a permissible number different from that of Paul, he will win if, when he throws again, he throws his chance before throwing that of Paul, and he will lose if he throws Paul's number before throwing his own.

---

* $q = 4$, $b = 2$, $B = 3$, $p = 13$, $c = 1$, $C = 1$ and
$^{52}C_7 = 4.13.17.10.7.47.46.$

(He is definite here about the number of dice being two, and this appears to be the number used, except in Italy. The references in Dante and Galileo's problem suggest that Hazard may have been played in Italy with three dice.*) Montmort gives the chances of Peter and Paul according to the rules he has laid down and then describes another game, which he says has no name and so he dubs it the game of Hope, and gives some calculations on this also. Backgammon however rather defeats him. He doesn't bother with the rules (they must have been entirely established by this time) and he calculates several simple chances but remarks that in the majority of situations the solution cannot be found. Remembering the intricacies of the game, one is inclined to agree with him. Apart from further complicated calculations on games involving 2, 3, 4, 5, 6, 7, . . . dice, in which the principles of calculating a conditional probability have been already laid down, using his cumbersome combinatorial notation, he does not appear to achieve anything else which is new. From an historical point of view, however, there is interest in his game of Nuts.

I have remarked earlier that divination among primitive tribes is (and was) carried out by casting pebbles, grain, or nuts, etc. It is also still a puzzle that the same ritual of divination was used in games to while away the idle hour. That this duality of purpose was probably universal, not just European, appears likely from Montmort's discussion on *Problème sur le Jeu des Sauvages*, *appellé Jeu des Noyaux*. He writes :

> Baron Hontan mentions this game in the second book on his travels in Canada, p. 113. This is how he explains it.
> It is played with eight nuts black on one side and white on the other. The nuts are thrown in the air. If the number of black is odd, he who has thrown the nuts wins the other gambler's stake. If they are all black or all white he wins double stakes, and outside these two cases he loses his stake.

His exposition of the chances involved is quite clear (he just refers

---

* The game of Three Dice as described by Montmort bears a distinct resemblance to his game of Hazard.

back to the Arithmetic Triangle), with the advantage to the
nut-thrower of 3/256. After some moral reflection he goes on :

> I think I should add that this problem was posed by me to a
> Lady, who gave me almost immediately the correct solution using
> the Arithmetic Triangle. But this table is useful only by chance,
> for if the nuts, instead of having two faces, had more than that,
> say four, this table would not be useful, and the problem would be
> less easy than the preceding.*

Having solved the problem for four-sided nuts he concludes his
book with Huygens' five problems, and some reflections on the
games of Her, Ferme and Tas. It is fairly clear that Montmort
modelled his book on what he thought was the plan followed by
James Bernoulli. For he writes :

> If I was going to follow M. Bernoulli's project I should have
> added a fourth part where I applied the methods contained in the
> first three parts to political, economic, and moral problems. What
> has prevented me is that I do not know where to find the theories
> based on factual information which would allow me to pursue my
> researches. . . . To speak exactly nothing depends on chance ;
> when one studies nature one is soon convinced that its CREATOR
> moves in a smooth, uniform way which bears the stamp of infinite
> wisdom and prescience. Thus to attach to the word " chance "
> a meaning which conforms with true philosophy one must think
> all things are regulated according to certain laws, those which we
> think dependent on chance being those for which the natural
> cause is hidden from us. Only after such a definition can one say
> that the life of man is a game where chance reigns.

Having reached this point he decides that while the rules of
probability can be applied to the game of life, the chances of
this game are too difficult to compute, much as it is too difficult
to compute the value of a throw in backgammon.

---

* Two reflections occur to me. It would be interesting to know who the
noblewoman was who had a facility in applying the Arithmetic Triangle
to games of chance. The principles of the calculation of probabilities must
have been generally known among educated persons.

The reasons and the different motives that men are able to have in order to move one way rather than another make it difficult to find out how they will act. Often they do not know where their own interest lies. . . . Caprice serves them rather than wisdom.

M. de Montmort obviously has very little idea of what James Bernoulli was going to put in the fifth part of the *Ars Conjectandi*.

The first edition of the *Essai d'Analyse* was published in 1708. In 1709 Nicholas Bernoulli presented his thesis *De Arte Conjectandi in Jure* for the degree of doctor of law at Basel University and then set off on his travels. When he went to Paris he met and became friendly with Montmort, staying with him on his country estate for some three months and afterwards keeping up a long correspondence with him. It was through Nicholas perhaps that Montmort also corresponded with John Bernoulli, James' brother and Nicholas' uncle, who had succeeded James in the chair of mathematics at Basel. This correspondence was also friendly in character, and on the whole Montmort seems to have been a likeable person. He had a wide circle of correspondents among mathematicians of all countries, including Newton and Leibnitz, exchanging with them news about mathematical problems and discussing solutions to the problems of the day (and fray). It is true that there is an echo of a stubborn temper in his running away from home at the age of 18, but this did not reappear in his later years. The greater part of his mathematical work was done at his house in the country. He appears to have been capable of great concentration since Malebranche recounted how he sat working with people playing the clavichord while his sons ran about the room and teased him. He was said to be quick-tempered and given to short, sharp bursts of anger, but it seems that because of his sweet nature he was very soon afterwards sorry and a little shamefaced. It is therefore more than a little strange that he reacted so sharply to de Moivre.

Francis Robartes, a fellow of the Royal Society of London and afterwards Earl of Radnor, was obviously interested in the calculation of probabilities since he wrote in 1693 a note "An Arithmetical Paradox, concerning the Chances of Lotteries" (*Phil. Trans.*, 17). In 1711 Abraham de Moivre wrote "De

Mensura Sortis, seu de Probabilitate Eventuum in Ludis a Casu Fortuito Pendentibus " which was also published in the *Philosophical Transactions* (**27**) and was afterwards expanded a little to form the first edition of the *Doctrine of Chances*. It was to this *De Mensura Sortis* that Montmort reacted so sharply. De Moivre wrote about this first memoir :

> The occasion of my undertaking this subject was chiefly owing to the Desire and Encouragement of the Honourable Francis Robartes, . . . who, upon occasion of a French tract, called *l'Analyse des Jeux de Hasard*, which had lately been published, was pleased to propose to me some Problems of much greater difficulty than any he had found in that Book ; which having solved to his satisfaction, he engaged me to methodise these Problems, and to lay down the Rules which had led me to their Solution. . . . Huygens, first, as I know, set down rules for the solution of the same kind of problem as those which the new French author illustrates freely with diverse examples.   But these famous men do not seem to have been accustomed to that simplicity and generality which the nature of the thing demands.   For while they talk of many unknown quantities so that they may represent various conditions of play, they set out their intricate calculations too meagrely, and while they always use equal skill in games, they contain this theory of games between much too narrow limits. . . .

The words from " Huygens " onwards are translated by the present writer from the Latin of *De Mensura Sortis* and they are not fair comment.   De Moivre's first attempts were largely derivative from this first edition of Montmort, and they do not, on the whole, display the masterly treatment of the subject which is evinced in the *Miscellanea Analytica* of 1730 or the third edition of the *Doctrine of Chances* (1756).   We note these words of de Moivre because he behaved in an unfair way to Montmort and gave him provocation to retaliate.   What is so surprising is that Montmort reacted so strongly.

At this time the English, commanded by Marlborough and under the shadow of the Grand Alliance, were marching about Europe defeating the French armies, and in 1708 and 1709 they were knocking at the gates of France.   It has been said of this period that the sciences were never at war, but the mystic attachment of every Frenchman for the soil of France may have caused in Montmort a reaction against the émigré Frenchman in London, especially when he was given just cause. It may also be said that one generally reacts most violently to one's own faults displayed in others, and it is possible that Montmort was not entirely easy in his conscience about his own publication ; for it may have been the thought that the *Ars Conjectandi* was unpublished at the time of issue of the *Essai d'Analyse*, and that he himself had learnt much of James' work from Nicholas, which caused him to resent de Moivre, The *Ars Conjectandi*, still unfinished, was at last published in 1713. eight years after the death of James, and it was followed in 1714 by the second edition of the *Essai d'Analyse*. The preface of the first edition was repeated, followed by an *Avertissement* in which Montmort wrote :

> The author did me the honour of sending me a copy. . . . M. Moivre was right to think I would need his book to reply to the criticism he made of mine in his introduction. His praiseworthy intention of boosting and increasing the value of his work has led him to disparage mine and to deny my methods the merit of novelty. As he imagined he could attack me without giving me reason for complaint against him, I think I can reply to him without giving him cause to complain against me. . . .

And reply he does over many pages, setting out the history from Pascal and Fermat onwards and always in what might be described as polemical fashion. This sort of behaviour is foreign to Montmort and it is, as far as I know, the only time in which he so indulged himself. Usually he carefully took the middle road in argument, as may be seen in his careful neutrality at the time of the calculus controversy. On one side was the army of English mathematicians and on the other John Bernoulli, with Leibnitz in the background. Each side tried to get Montmort to join them, but his philosophical and ecclesiastical training stood him in good stead and he managed to avoid getting embroiled. On a visit to England in 1715 he became reconciled to de Moivre and was made a Fellow of the Royal Society. In Paris in 1716 he was made a member of the Académie des Sciences. He frequently visited Paris for business reasons and on the last of these trips he caught smallpox and died of it on October 7th, 1719.

The second edition of the *Essai d'Analyse* is more than twice as long as the first and reflects to a certain extent the maturity of thought on the subject which came to the author, possibly as a result of his conversations and exchange of letters with Nicholas Bernoulli. The history in the *Avertissement* is interesting if only because it indicates how little was known of the story of the development of the calculation of chances. According to Montmort, Pascal originated most of combinatory theory and Pascal and Fermat the calculation of chances. He gives a clear review of Huygens' and James Bernoulli's work, with many eulogistic references also to Leibnitz. (If Montmort is a fair example, and in this instance he probably is, Leibnitz seems to have been greatly venerated by the continental mathematicians.) Possibly because he has been somewhat scathing about de Moivre and English mathematics he gives some space to a memoir by Craig, which is interesting only because of later French work on the credibility of witnesses. Montmort does not, I think, know how to take this memoir. He writes to begin with :

> I have found in the *Phil. Trans.* a memoir in which it is proposed to estimate the probability of men speaking the truth, whether in speech or in writing : but can one find this? If an emphatic yes gives a semblance of truth a/b, an emphatic yes of an emphatic yes will give a semblance $a/b.c/d$ of truth if the witness of the second is not of the same strength, and a^2/b^2 if they hold the same authority, . . . which is obvious. But what can one conclude from this, and how can one apply these theories? I think that this is impossible. It has however been undertaken . . . by an English mathematician. . . . The book of which I speak has for title *Philosophiae Christianae Principia Mathematica*. M. Craig is the author. . . . The Author tries in the main to prove, against the Jews, the story of Jesus Christ and to demonstrate to libertines that the choice they have made in preferring the pleasures of this world . . . to the expectation of those . . . who follow the law of the Evangelist, is not reasonable and does not accord with their true interests. . . .

In spite of his not believing that Craig's calculations are possible he feels that they are of considerable theoretical interest.

For myself I find the design of M. Craig pious and worthy of praise, and the execution of it as good as it can be, but I believe this work much more suitable as an exercise for mathematicians than as a means to convert the Jews or the incredulous. One can certainly conclude after reading this treatise that the Author is very ingenious, that he is a great mathematician and highly intelligent. The clarity of mathematics and the saintly obscurity of the faith are two entirely opposing things : I do not think that anyone will ever succeed in combining them.

One wonders a little why James Bernoulli's projected fifth part of the *Ars Conjectandi* should have been considered practicable whereas Mr. Craig's design was not. The remainder of the *Avertissement* is principally concerned with the story of the calculation of chances during the years immediately preceding his own writing, and although more or less accurate it is possibly unduly flattering to Leibnitz.

The main body of the second edition contains a great deal of new material and has incorporated at the end letters which passed between the Bernoullis, Nicholas and John, and Montmort. It is clear that the Bernoullis helped considerably with this second edition, clarifying Montmort's ideas for him and contributing much in the way of summation of series. James was very good at summing series, so that this type of mathematical exercise was easy for Nicholas and John. Todhunter gives a clear, full account of the majority of the mathematics, and there is little to add, but to make the story complete we repeat a little of what he has already described in detail. The theorems on combinations are, in the second edition, brought forward to the beginning of the treatise. This, in a way, is symbolic of the change of treatment throughout the book. Montmort has come to maturity and discusses, wherever he is able, the general solution to the problem he sets himself, rather than beginning with special cases and then plunging into a general statement which is often unsupported by mathematical argument. He still retains his combinatorial notation and introduces the symbol $\boxed{q}$ for the qth figurate number. The expansion involved in the solution of his Proposition XVI,

> Throwing randomly any number whatever, d, of dice, of which the number of faces, f, may be also whatever one wishes, to find what is the number of ways of throwing a given number p,

illustrates both his general method of attack and introduces his new notation. In the first edition he gives, presumably by exhaustive enumeration, the chances involved in throwing with six-sided dice the numbers $1, 2(1), \ldots, 8$. Now he starts off:

> Let $p - d + 1 = q$, and denote by the arbitrary symbol $\boxed{}$ the figurate number of order d, which corresponds to q; this means the first number of order d if $q = 1$, and the second of order d if $q = 2$ and the third if $q = 3$, etc. The formula

$$\boxed{q} - d \times \boxed{q - f} + \frac{d.d - 1}{1.2} \times \boxed{q - 2f} -$$

$$- \frac{d.d - 1.d - 2}{1.2.3} \boxed{q - 3f} + \frac{d.d - 1.\, d - 2.\, d - 3}{1.2.3.4} \boxed{q - 4f} -$$

etc. will express the looked-for number.

This formula, the differences of zero series, had been reached by de Moivre in *De Mensura Sortis* in 1711. It is a generalisation of Proposition XXXI of the first edition, but Montmort had no idea then (1708) as to how it might be solved generally. De Moivre probably took the problem from the first edition, generalised it, and gave the solution with no indication of method of proof. It is just possible that he was following Montmort's own method and guessed the general result from the particular cases, but in the light of his undoubted analytic powers this is unlikely. Montmort reached the solution by himself also, since a letter to John Bernoulli in 1710 shows that he had already obtained it.

The generalisations of the various topics discussed in the first edition are interesting, without adding anything particularly new to the probability calculus, although the various methods for the summation of series show the skill of the Bernoullis in that part of algebra. The matching distribution is presented with a proof of the general case ; this proof was not given in the first edition, implying that Montmort either guessed the original solution or was dissatisfied with his first method of proof. The proof which he now gives is due to Nicholas Bernoulli, but he repeats his own exponential limiting form. The Problem of Points solved in full generality with two players of unequal skill is presented again with the help of Nicholas. I shall consider this in the discussion of the *Doctrine of Chances* since the problem did not cease to be of interest for many years. There are the first fumblings towards the questions of annuities, and the analysis of the chances in one or two further games of chance such as Her, not previously given. The most interesting addition is however the printing in full of the letters between himself and the Bernoullis ; these are at the end of the book and occupy over one hundred pages. They begin after Nicholas had left Paris and returned to Basel. John writes in a lofty way and obviously enjoys pointing out that Montmort had missed writing down the sum of a geometric progression on at least two occasions. He does not show any great capacity as a probabilist. Both John and Nicholas take as their focus of comment the first edition of the *Essai d'Analyse*, but it is Nicholas who comes forward with helpful and sometimes new solutions. Montmort obviously published the long series of letters because he wanted Nicholas to have the credit of the results he had worked out. Much of what Nicholas did was a generalisation of problems which had been proposed by Montmort or his uncle James. He does, however, also set out the problem which later was to become famous as the St. Petersburg problem, and he also sets out his uncle's "golden theorem" as if it were his own, adding " I recall that my uncle has demonstrated a similar thing in his treatise *Ars Conjectandi* now being printed at Basel". Nicholas was an excellent mathematician, with the algebraic diversity and versatility which could be expected from anyone of ability who had been trained in part by James. One may be pardoned perhaps for wondering whether many of his ideas were not derived from the same source.

On September 5th, 1712, Montmort wrote to Nicholas a tirade about *De Mensura Sortis* which he had just finished reading. He is seething about it because there is nothing new in it if one takes into account the letters between himself and Nicholas, which were not yet published, as well as the first edition of the *Essai d'Analyse*.

> ... You will find that the problems he discusses, which are not solved, are solved in our letters. Moreover I do not think there is in this work, elsewhere very good, anything new to you, and nothing which will give you pleasure by its originality. . . .

The letter, of extreme length, is that of a very angry man. Nicholas, contrary to the usual Bernoulli practice of joining in battles, here attempts to soothe his friend. He wrote from London on October 11th, 1712, before he had received Montmort's outburst, that

> I have had the pleasure of often seeing here M. de Moivre who has given me a copy of his book *De Mensura Sortis*. He tells me that he has also sent you a copy and that he is awaiting with impatience your views on his work. You will be surprised to find there many of the problems which we have solved. . . .

Nicholas wrote again from Brussels at the end of 1712 (December 30th) explaining that Montmort's letter had had to follow him to London and then to the Low Countries and had consequently been delayed. He dealt with Montmort's letter in detail prefacing his remarks with

> I am content that you have received M. de Moivre's book *De Mensura Sortis*. It is true that nearly all the problems proposed there have been solved either in your book or in our letters. As I knew that M. de Moivre awaits with impatience your judgement on his book, I have taken the liberty of sending him the substance of your remarks. . . . *

---

* Apart from de Moivre's correspondence with John Bernoulli, which has nothing to do with probability, there are no letters written by de Moivre known to the present writer and a search has so far failed to find any. Yet it is not beyond the bounds of possibility that de Moivre wrote to Montmort after receiving a letter from Nicholas. Montmort would probably have been too angry to answer. Todhunter states that there *was* a correspondence between Montmort and de Moivre, but I have not been able to trace it.

Nicholas is soothing to Montmort but fair to de Moivre, pointing out in several places in his commentary that de Moivre had shown to him his general solutions to various problems when he was in London, and trying to explain that de Moivre had not intended to slight Montmort by his introduction.*

> I do not know if M. de Moivre has had the intention in his preface of insulting you as you think : for myself I hold that the methods you have given in your book are good enough to solve all the general problems of M. Moivre, which for the most part differ from yours only in the generality of the algebraic expression. As I am persuaded that M. Moivre himself would do you the justice of recognising that you have taken this subject much farther than M. Huygens and M. Pascal, who have given only the first elements of the science of chance, and that after them you have been the first who has published general methods for these calculations. . . .

(One wonders a little about James Bernoulli.) That Montmort was not appeased is apparent from the *Avertissement* on which I have commented. Apart from a discussion by Nicholas on the problem of the parity of the proportion of male and female births, which was being discussed in London at the time of his visit, there is little more of interest in the correspondence, which appears to have finished in 1713 just before the publication of the second edition of the *Essai d'Analyse*.

With the publication of this second edition Montmort seems to have given up researches on the probability calculus. It may have been that the short history which he wrote about the theory of probability (or possibly the calculus controversy) piqued his curiosity, but he wrote to Nicholas (August 20th, 1713) :

> I would very much like to know what you and your uncle think of the book entitled *Commercium Epistolicum*, etc., that the Royal Society has had printed to assure to M. Newton the glory of having been the first and only one to have invented the new methods. I promise you secrecy if you tell me. Everyone here waits M. Leibnitz's answer.

---

* I find this a little difficult to believe.

It is to be desired that someone would take the trouble of instructing us how and in what order the discoveries in mathematics have succeeded one another and to whom we are obliged for them. We have a history of painting, of music, of medicine, etc. A history of mathematics, and in particular of geometry, would be a very useful work. What pleasure would one not have to see the link-up, the connection between the methods, the relation between the different theories, beginning from the first stirrings up to the present?

Whatever started him off, he undertook the longed-for work himself and at the time of his death was engaged in compiling a history of geometry. None of these manuscripts seem to have survived his death.*

Montmort's importance from the probability point of view is possibly not in the new ideas which he introduced but in the algebraic methods of attack. These were perhaps much the same as those of James Bernoulli, but the two mathematicians, coupled with Nicholas, reinforce one another. They must have given inspiration to many other pure mathematicians, among them de Moivre, who would not have been interested solely in the laborious enumeration of the fundamental probability set.

References

Montmort does not seem to have been written about as much as he deserves. Hoeffer has little to contribute, citing only the *Éloge* by Fontenelle which latter I have used for biographical detail. The letters between himself and Nicholas and John Bernoulli also help to form opinion.

---

* Many early scientific manuscripts and letters seem to have gone astray at the time of the French Revolution. Some were dispersed, possibly for safety, among members of the Académie des Sciences, and have come to light at intervals since. The notebooks of Lavoisier, for instance, were retrieved only during this present century. It is not impossible that further discoveries of this kind await future investigators.

Abraham de Moivre—painted by Jos. Highmore, 1736

(*see page* 162)

Plate 9

chapter **15**

Abraham de Moivre and the " Doctrine of Chances "

Heureux qui, comme Ulysse, a fait un beau voyage.

JOACHIM DU BELLAY

Abraham de Moivre was born at Vitry in Champagne on May 26th, 1667. His father is described as being a surgeon, a Protestant, and neither wealthy nor of noble birth.* At the age of eleven he was sent to the Protestant college at Sedan and, when that was suppressed, to the college at Saumur. Both at Sedan and at Saumur he studied the humanities, and it was not until he went to Paris and came into contact with the great Ozanam† that he became interested in mathematics, although he is stated to have read Huygens' *De Ratiociniis in Aleae Ludo* while at Saumur.‡ De Moivre's family moved to Paris, which is possibly why he left Saumur to go to the Sorbonne. This was, however, his good fortune since Ozanam was one of the great teachers of mathematics. Under him de Moivre studied geometry, trigonometry, mechanics, perspective, spherical trigonometry, Euclid and the spherics of Theodosius. On October 18th, 1685, the Edict of Nantes was revoked and de Moivre, then eighteen, was imprisoned in the Priory of St. Martin. There is no record of what happened to his family and his immediate relations. All that is

---

* It has been suggested that Moivre added the noble prefix " de " while crossing the Channel. Since he is usually known as *de* Moivre I have followed current practice.

† Jacques Ozanam (1640–1717), born at Bouligneux (Aisne).

‡ The usual story is told of how he stumbled upon Euclid's propositions. This is the same story with variations that was told about Blaise Pascal and James Bernoulli (among others).

certain is that (in his twenty-first year) on April 27th, 1688, he
was set free and that he left France immediately. He never
returned to France and never published anything in French.
Early habits persisted, however, to the extent of correspondence :
such few letters of his as are known to exist are written in French.
Highmore's portrait (see Plate 9, facing page 160) shows him
when approaching his seventieth year.

De Moivre, then, at the age of twenty, cut himself off completely from his native country and landed in England without
money, friends or influence, his only capital being his knowledge of mathematics. This, since he had been taught by Ozanam,
he believed to be profound, but even here he was disillusioned.
He set up a school in Fleet Street, London, and gradually worked
up a connection as visiting tutor to the sons of several noblemen.
Newton presented a copy of the *Principia Mathematica* to the Earl
of Devonshire. While tutoring at the Earl's house, so the story
runs, de Moivre saw this copy and found that it was beyond him.
He is said to have bought a copy of his own, to have torn it into
separate pages, and to have learned it page by page as he walked
London from one tutoring job to another.

Presumably he spent his first years in England teaching and
making the acquaintance of English mathematicians. It is not
known how, but possibly by an introduction from Ozanam, that
in 1692 he met Edmund Halley, then Secretary to the Royal
Society. He became friendly with him and in so doing he broke
into the charmed circle of mathematical nobility. This friendship
with Halley possibly inspired the young Frenchman to mathematical research, for in 1695 Halley presented for him his first
paper (on fluxions) to the Royal Society. The paper does not
seem to have been of much consequence. Halley, the astronomer,
tried, it is said, to get de Moivre to interest himself in astronomy
but without success. It is not, however, too far-fetched to see a
link between Halley's construction of a life-table* and de Moivre's
life-long interest in the theory of annuities. De Moivre was elected

* " An estimate of the Degrees of Mortality of Mankind, drawn from curious
Tables of the Births and Funerals at the City of Breslau, with an Attempt
to ascertain the Price of Annuities upon Lives." *Phil. Trans.*, 1693.

a Fellow of the Royal Society in 1697. It is unlikely that he met Newton before this, since Newton was at that time still at Cambridge and possibly they did not become friendly until after 1703 when Newton moved to London to become Master of the Mint, and shortly afterwards President of the Royal Society. According to the Éloge of de Moivre (written by Grandjean de Fouchy), de Moivre was in the habit of going to a coffee-house after his day's tutoring and schooling, and where, presumably, he began to augment his income by calculating odds for gamblers. Newton occasionally called at the coffee-house and carried de Moivre back to his house for talk. In the first edition of the *Doctrine of Chances* (1718), which was dedicated to Newton, de Moivre wrote :

> The greatest help I have received in writing on this subject having been from your incomparable works, especially your method of series, I think it my duty publicly to acknowledge that the improvements I have made in the matter here treated of are principally derived from yourself. The great benefit which has accrued to me in this respect, requires my share in the general tribute of thanks due to you from the learned world. But one advantage, which is more particularly my own, is the honour I have frequently had of being admitted to your private conversation, wherein the doubts I have had upon any subject relating to Mathematics have been resolved by you with the greatest humanity and condescension. Those marks of your favour are the more valuable to me because I had no other pretence to them, but the earnest desire of understanding your sublime and universally useful speculations. . . .

As Master of the Mint Newton was no longer so urgently interested in mathematical exposition and, it is related, when approached by earnest students he would say " Go to Mr. de Moivre ; he knows these things better than I do." This friendship with Newton appears to have lasted until the latter's death in 1727.

These first ten years of the eighteenth century saw de Moivre growing in intellectual stature and gradually becoming on terms of friendship with most of the famous mathematicians of the day. On the whole he seems to have been a friendly person, but patience and forbearance may have been thrust upon him by his humble circumstances. There is a certain defect of character somewhere, for none of his friends exerted their influence on his behalf, although it would probably have been easy for them to do so. The English are adept at trimming their sails to catch the prevailing wind—taking the middle path it is euphemistically called—but they are suspicious of any outsider who tries to do the same, and de Moivre was not free from a suggestion of this in several controversies. However, in the controversy with George Cheyne he himself was attacked, and because John Bernoulli was also, this wrought a friendship between them. George Cheyne, a Scots medical who is said at one time in his life to have weighed 448 lbs, wrote (at the age of thirty-two) a tract entitled *Fluxionum methodus inversa sive quantitatum fluentium leges generaliores* in which he asserted his priority in a number of theorems of the new calculus and made attacks on various mathematicians of the day. Seen from the twentieth century this would appear ephemeral, but de Moivre possibly felt himself not yet firmly established and so he wrote a reply to it in 1704, *Animadversiones in D.G. Cheynaei tractatum de fluxionum methodo inversa*. Both John Bernoulli and de Moivre wrote to Cheyne and wrote to each other. John Bernoulli pointed out to Cheyne and to de Moivre the mathematical errors which had been made; Cheyne then published the corrections without acknowledgement in 1704 in *Addenda et adnotanda in D. G. Cheynaei libro* before the *Animadversiones* of de Moivre had appeared. Neither de Moivre nor John Bernoulli made any reply to this second paper of Cheyne, although they pointed out to each other the errors of Cheyne's methods and manners. The correspondence between de Moivre and John Bernoulli, written in French with an occasional long tag in Latin, are of the usual scientific chitchat of the day. There is on the whole little of factual importance in them, but they do illuminate the characters of the two men. De Moivre was obviously dissatisfied with his lot, understandably so. He begged John to intercede with Leibnitz to use his influence to get him a university post somewhere.

> . . . he (Leibnitz) is so universally respected, and one has too much confidence in him, that a word from his mouth in my favour will be extremely advantageous. . . .

He went on to describe how he had to tramp about the city from pupil to pupil, and how he was not able to do enough work to enable him to save money.

> If there were some position where I could live tranquilly and where I could save something, I should accept it with all my heart. . . .

But nothing ever came his way.

In 1711 de Moivre published *De Mensura Sortis, seu de Probabilitate Eventuum in Ludis a Casu Fortuito Pendentibus* which, as has been noted, led to difficulties with Montmort. Montmort objected, rightly, to de Moivre insinuating that he had done no more than improve on Huygens. De Moivre objected to Montmort's insinuation that he had taken his ideas from the *Essai d'Analyse*. Both were partially right, but I think Montmort had a grievance. There is little evidence that de Moivre had thought about games of chance with the idea of analysing probabilities with the full weight and development of the (then) modern algebra until after the publication of Montmort's book.* It was therefore ungracious of de Moivre to write as though Montmort had added nothing to the mathematical approach to probability, and it was, of course, not too difficult for a mathematician of de Moivre's capacity to generalise Montmort a little and to invent new methods of attack. The vigour of Montmort's reply to the *De Mensura Sortis* in the *Avertissement* of the second edition of *Essai d'Analyse* published in 1714 has been noted. By 1715, however, the quarrel seems to have resolved itself.† Montmort visited London in that year and de Moivre acted as his interpreter during that visit. In 1718 the first edition of the *Doctrine of Chances* was published by de Moivre. It is an enlarged version of *De Mensura Sortis*, written in

---

* Montmort likewise probably had no such ideas until he read James Bernoulli's papers and the summary of his book.

† The fact that the quarrel was so quickly over makes the idea that de Moivre and Montmort corresponded, possibly through Nicholas Bernoulli, not too fanciful. I have not been able to trace any such correspondence, but it may just be a question of looking in the right place for it. That Montmort's manuscripts disappeared after his death does not, however, make the quest a hopeful one.

English, with arithmetical illustrations added for those gentlemen
who do not understand algebra. In the Preface we read :

> I flatter myself that those who are acquainted with Arithmetical
> Operations, will, by the help of the Introduction alone, be able to
> solve a great Variety of Questions depending on chance....I have,
> as much as possible, endeavoured to deduce from the Algebraic
> Calculations several practical rules, the Truth of which may be
> depended upon, and which may be very useful to those who have
> contented themselves to learn only common Arithmetick. . . . I
> must take notice to such of my Readers as are well versed in
> Vulgar Arithmetick, that it would not be difficult for them to
> make themselves Masters, not only of the Practical Rules in this
> Book, but also of more useful Discoveries if they would take the
> small Pains of being acquainted with the bare Notation of Algebra,
> which might be done in the hundredth part of the Time that is
> spent in learning to read Short-Hand. . . .

Thus the tone of the book is set, and the author keeps to it
throughout. There is, however, no disguising the fact that he
has been interested in the algebra and that he finds it necessary
to explain the provenance of the book. He tells us that he started
work on games of chance at the entreaty of the Hon. Francis
Robartes, and that when he wrote *De Mensura Sortis* he had read
nothing but Huygens' tract. In order to excuse his remarks about
the *Essai d'Analyse* he writes :

> As for the French book, I had run it over but cursorily, by
> reason I had observed that the Author chiefly insisted on the Method
> of Huygens, which I was absolutely resolved to reject. . . . However,
> had I allowed myself a little more time to consider it, I had certainly
> done the Justice to its Author, to have owned that he had not only
> illustrated Huygens' Method by a great variety of well chosen
> examples, but that he had added to it several curious things of his
> own Invention. . . . Since the printing of my Specimen, Mr. de
> Montmort, Author of the *Analyse des Jeux de Hazard*, published a
> Second Edition of that Book, in which he has particularly given
> many proofs of his singular Genius and extraordinary Capacity ;
> which Testimony I give both to Truth, and to the Friendship with
> which he is pleased to Honour me. . . .

The words after " reject " were dropped from the second and third editions (1738 and 1756). These were published after the death of Montmort (1719), and de Moivre, in this the first edition, was undoubtedly trying to curry favour with Montmort and with the Bernoullis.

This preface to the first edition is remarkable for its plain common sense. There are no appeals to the Deity but a reliance on combinatorial theory. About ill-luck he writes :

> The asserters of Luck are very sure from their own experience that at some times they have been very Lucky, and that at other times they have had a prodigious run of ill luck against them. . . . They would be glad for instance to be Satisfied, how they could lose fifteen games together at Piquet, if ill Luck had not strangely prevailed against them. But if they will be pleased to consider the Rules delivered in this Book, they will see that tho' the Odds against their losing so many times together be very great, viz. 32767 to 1, yet that the Possibility of it is not destroyed by the greatness of the Odds, there being One Chance in 32768 that it may so happen, from whence it follows that it was still possible to come to pass without the intervention of what they call ill luck....*

There is also some idea of using probabilities to judge between two alternatives, although unfortunately he does not develop this very far. Thus we are told :

> The same arguments which explode the Notion of Luck may, on the other side, be useful in some Cases to establish a due comparison between Chance and Design. We may imagine Chance and Design to be as it were in Competition with each other for the production of some sorts of Events, and may calculate what Probability there is, that those Events should be rather owing to one than to the other. . . . From this last Consideration we may learn in many Cases how to distinguish the Events which are the effect of Chance, from those which are produced by Design.

He also remarks, in discussing the Problem of Points, when the two gamblers are of unequal skill, that

---

* c.f. R.A. Fisher's *The Design of Experiments*.

It is true that this degree of skill is not to be known any other way than from Observation : but if the same Observation constantly recur, 'tis strongly to be presumed that a near Estimation of it may be made. . . .

In addition to telling the reader the author's ideas about chance, the Preface serves as a prolonged summary of the contents of the book, and these contents are undoubtedly worthy of comment. The writing is didactic and one might deduce from it that de Moivre was a teacher since the style is that of a text-book. He begins by setting out the definition of probability, of the addition of probabilities, of expectation, of the independence of events, of joint probabilities, of conditional probabilities. He is also quite definite about the probability of *at least one* or *at least two* successes or, etc.

If the events in question are n in number, and are such as have the same number a of Chances by which they may happen, and likewise the same number b of Chances by which they may fail, raise $a + b$ to the power n. And if A and B play together, on condition that if either one or more of the Events in question do happen, A shall win and B lose, the probability of A's winning will be $\dfrac{(a + b)^n - b^n}{(a + b)^n}$ and that of B's winning will be $\dfrac{b^n}{(a + b)^n}$ . . . .

He goes on to say that if the conditions of play are such that A wins if there are two or more happenings then A's chance of winning is
$$\frac{(a + b)^n - nab^{n-1} - b^n}{(a + b)^n}$$
and so on. Having laid down the framework of his rules he proceeds to " The Solution of Several Sorts of Problems deduced from the Rules laid down in the Introduction". The first four problems are trivial, requiring merely the substitution of numbers in his algebraic formulae. In Problem V he reaches what has been commonly called Poisson's* binomial exponential limit. Since, as we shall later note, the other approximation to the sum of a

* Siméon-Denis Poisson (1781–1840), born at Pithiviers.

number of binomial terms was also reached by de Moivre, it is of interest to consider this present exponential limit in full. I set it out verbatim.*

Problem V. To find in how many trials an event will probably happen, or how many trials will be requisite to lay on its happening or failing ; supposing that a is the number of chances for its happening in any one trial and b the number of chances of its failing.

Solution.

Let x be the number of trials ; therefore, by what has been already demonstrated in the Introduction, $(a + b)^x - b^x = b^x$, or $(a + b)^x = 2b^x$; therefore $x = \dfrac{\log 2}{\log(a + b) - \log b}$. Moreover let us reassume the Equation $(a + b)^x = 2b^x$ and making $a,b :: 1,q$ the equation will be changed into this $\left(1 + \dfrac{1}{q}\right)^x = 2$; let therefore $1 + \dfrac{1}{q}$ be raised actually to the power x by Sir Isaac Newton's theorem and the equation will be

$$1 + \frac{x}{q} + \frac{x(x - 1)}{1.2.q^2} + \frac{x(x - 1)(x - 2)}{1.2.3.q^3} \text{ etc.} = 2.$$

In this equation if $q = 1$ then will x likewise be $= 1$; if q be infinite then will x also be infinite. Suppose q infinite, then the equation will be reduced to $1 + \dfrac{x}{q} + \dfrac{x^2}{2q^2} + \dfrac{x^3}{6q^3}$ etc. $= 2$. But the first part of this Equation is the number whose hyperbolic logarithm is $\dfrac{x}{q}$, therefore $\dfrac{x}{q} = \log 2$.

In the third edition (1756) he expands $\log \left(1 + \dfrac{1}{q}\right)$ in the equation $x \log \left(1 + \dfrac{1}{q}\right) = \log 2$ and neglects all terms but the first. Given the method of procedure, i.e. letting x and q tend to

---

* I have retained de Moivre's notation everywhere.

infinity but x/q remaining finite, it would not be difficult for the mathematicians who came after de Moivre to generalise his work. Problem V is followed by a lemma :

> To find how many chances there are upon any number of dice, each of them of the same given number of faces, to throw any given number of points.

The demonstrated solution to this is the well-known expansion for the difference quotients of zero which was first given ever in *De Mensura Sortis*. Problem VIII gives us the multinomial.

> Three gamesters A, B, C play together on this condition, that he shall win the set who has soonest got a number of games ; the proportion of the chances which each of them has to get any one game assigned, or, which is the same thing, the proportion of their skill being respectively a, b, c. Now after they have played some time, they find themselves in this circumstance, that A wants one game of being up, B two games, and C three ; the whole stake between them being supposed 1. What is the expectation of each?

The method used is to take $(a + b + c)^4$, expand, and sort out the terms. (It is also the method arrived at independently by Montmort and Nicholas Bernoulli in their correspondence.) At the end we have

> If n be the number of all the games that are wanting, p the number of gamesters, and a, b, c, d, etc. the proportion of the chances which each gamester has respectively to win any one game assigned ; let $a + b + c + d$ etc., be raised to the power $n + 1 - p$, then proceed as before.

A simple gambler's ruin problem follows, with a demonstration of how to calculate the probabilities by using logarithms if the numbers involved are large, then a problem involving conditional probabilities with an exposition of how to sum a series by differencing the terms until constant differences are obtained. The problems on Bassette and Pharaoh present nothing new, I think, although the demonstrations of how to sum the series involved are remarkable for their clarity. The emphasis on the summing of series and the method of detached coefficients (which he invented for himself) possibly arose from talks with Nicholas Bernoulli and with Montmort. The algebraic attack adopted by de Moivre would, inevitably, necessitate his having to find such sums, and Montmort revealed his own deficiencies in this analysis by publishing the letters from John Bernoulli in which he pointed this out. De Moivre shows by his discussion that he is an analytic master of great power, although he can on occasion go astray. His approximation to the differences of zero series, the idea for which he attributes to Halley, will be found to be extremely poor for large numbers. The remainder of the first edition is devoted to problems relating to the duration of play. These problems are thoroughly worked over with the author's usual ingenuity and clarity, but they do not present anything in the way of new probability principles. One of his methods indicates that he must already have thought of the expansion of cos $n\,\theta$ in terms of cos θ, an expansion now known as de Moivre's Theorem, but he does not state it explicitly. Whether de Moivre gets the credit for many of his discoveries and innovations is immaterial. There is no doubt whatsoever that this, the first edition of the *Doctrine of Chances*, is written by a man who was already superior to Montmort and the Bernoullis in his mathematical powers, and who, when he came to maturity, was to produce in his third edition the first modern book on probability theory.

The beginning years of the eighteenth century were enlivened for mathematicians by the controversy about the origin of the new calculus of fluxions. De Moivre wrote to John Bernoulli in 1705 that he was not taking sides in the controversy and he added some unusually flowery compliments about the skill of James and of John in their applications of this calculus. None the less in 1711 he agreed to serve on the committee set up by the Royal Society which was to judge priorities. It is possible that this led to the correspondence between John and himself petering out, although in his last letter John writes* :

---

* This is a strange remark considering that John was a fervent champion of Leibnitz and carried on the controversy after Leibnitz's death in 1716.

I am glad you do not suspect me of having taken the side of M. Leibnitz in the quarrel between him and M. Newton. . . .

There were three editions of the *Doctrine of Chances*—in 1718, 1738 and a posthumous one in 1756. The *Miscellanea Analytica de Seriebus et Quadraturis*, important in that it contains the first attempt at Stirling's theorem, was published in London in 1730. The other great work of de Moivre's, *Annuities on Lives*, editions of which were published in 1724, 1743 and 1750, with an Italian translation by P. Fontana in 1776, embroiled him in a controversy with Thomas Simpson. Simpson published *The Nature and the Laws of Chance* in 1740 ; this is substantially the same in content as the 1738 edition of the *Doctrine of Chances*. In 1742 he did the same thing with a treatise on annuities, which possibly prompted de Moivre's reissue of his own work in 1743 with a protest about Simpson in the introduction. This plagiarism on the part of Simpson, for it was little more, may have been a serious matter for de Moivre, since he never had a settled position, was by now growing old—he was 76—and needed the money which he made from the sale of his books. De Moivre does not appear to have pursued the quarrel with Simpson and issued no reply to an attack on himself made by a friend of Simpson's in the *Penny Encyclopaedia* ; this attack was unjust and not representative of the facts. Simpson was not a pleasant character and it is justice that his one big contribution to probability in 1757, *An Attempt to show the Advantage arising by Taking the Mean of a Number of Observations in Practical Astronomy*, should have been plagiarised from him in his turn and the problems proposed by him adopted by Lagrange without acknowledgement.

De Moivre outlived all his friends, with the exception of James Stirling. He worked at mathematics all his life, for after the publication of the *Miscellanea Analytica* the applications of his mathematical analysis were incorporated in the subsequent editions of his books. He died on November 27th, 1754, at the age of 87, the cause of his death being recorded as from " somnolence". It is difficult to sum up the value of de Moivre's life and work from the point of view of probability because his writings have had such a tremendous impact on the theory. He was undoubtedly started off by Montmort and by James Bernoulli, and because he was a greater mathematician than either of them, and was spurred by poverty, he accomplished much more. But it is possible that if he had been the first of the three, rather than the last in time, he might have accomplished much less. If he declined to admit his debt to Montmort* he was always more than respectful in his references to James Bernoulli, even in the face of derogatory remarks about James by John. As an old man he wrote graciously about James, who had then been dead nearly fifty years. Yet his debt to James was possibly no greater than his debt to Montmort, and he owed to both of them at the very least the idea that algebra could be developed and extended and applied to games of chance. His timidity and lack of staying power in controversy is only equalled by his boldness and enterprise in mathematical attack. Thus for his solutions on the Duration of Play he gives us the first idea of generating functions for probabilities, while algebraically he finds it necessary to create the concept of recurring series and the expansion of cos $n\theta$. The third edition of the *Doctrine of Chances* follows the same pattern that was laid down in the first edition, with one or two significant additions, but here and there the mathematical argument is tightened, and the maturity of thought easily makes itself felt. It would seem that the *Miscellanea Analytica* (1730) summarises the results of (say) nearly thirty years of mathematical research, while the *Doctrine of Chances* (1756) summarises the results of half-a-century's thinking about the probability calculus, and it is likely that the latter was responsible for most of the mathematical invention.

The *Miscellanea Analytica* is interesting today to probabilists because it was here that de Moivre first gave the expansion of factorials which now goes by the name of Stirling's theorem.† Some of the copies printed in 1730 contain a Supplement in which he has further thoughts on the expansion of factorials. Sometime, very shortly after this 1730 printing, de Moivre had

---

* It is just possible that his unhappy early life gave him an unconscious prejudice against the French nobility.

† It is interesting throughout from a general point of view for its freshness of mathematical attack and for its creativeness.

still further thoughts which he committed to paper in 1733. Writing in 1754 the old man says :

> I shall here translate a Paper of mine which was printed November 12, 1733, and communicated to some Friends, but never yet made public, reserving to myself the right of enlarging my own thoughts, as occasion shall require.

At the end of his long life his memory may have been a little shaky, since the paper was in fact incorporated in some copies of the *Miscellanea Analytica** as a second Supplement. The paper (as the rest of the book) is in Latin, but it is certainly the same algebra as that given in the third edition of the *Doctrine of Chances*, pp. 243–254. Because it is of general interest a verbatim copy of the matter in the *Doctrine of Chances* is given as Appendix 5. It is entitled :

> A Method of approximating the Sum of the Terms of the binomial $(a + b)^n$ expanded into a Series, from whence are deduced some practical rules to estimate the Degree of Assent which is to be given to Experiments.

This method of approximating to the " Sum of the terms of the binomial " is the first " normal " limit to be derived in probability, since this " Paper of mine " of 1733 is just this.† There is no doubt that this approximation was sought after by de Moivre, and found in order to use James Bernoulli's " golden theorem ". He speaks of the work of James and of Nicholas in his introductory remarks and then proceeds :

> If the binomial $(1 + 1)$ be raised to a very high power denoted by n, the ratio which the middle term has to the sum of all the terms, that is to 2^n, may be expressed by the fraction
>
> $$\frac{2A\,(n-1)^n}{n^n\sqrt{n-1}}$$

---

* Karl Pearson suggested that the rarity of this second Supplement of the *Miscellanea Analytica* means that it was only incorporated in such copies as had not been sold by 1733. This is undoubtedly the right explanation.

† I find it curious that de Moivre did not incorporate it in the Second Edition of the *Doctrine of Chances* in 1738.

wherein A represents the number of which the hyperbolic logarithm is

$$\frac{1}{12} - \frac{1}{360} + \frac{1}{1260} - \frac{1}{1680}, \text{ etc.}$$

But because the quantity $\dfrac{(n-1)^n}{n^n}$ or $\left(1 - \dfrac{1}{n}\right)^n$ is very nearly given

when n is a high power, which is not difficult to prove, it follows that, in an infinite power, that quantity will be absolutely given, and represent the number of which the hyperbolic logarithm is — 1 ; from whence it follows, that if B denotes the number of which the hyperbolic logerithm is

$$-1 + \frac{1}{12} - \frac{1}{360} + \frac{1}{1260} - \frac{1}{1680} \text{ etc.}$$

the expression above written will become $\dfrac{2B}{\sqrt{n-1}}$ or barely $\dfrac{2B}{\sqrt{n}}$, **and**

that therefore if we change the signs of that series and now suppose that B represents the number of which the hyperbolic logarithm

is $1 - \dfrac{1}{12} + \dfrac{1}{360} - \dfrac{1}{1260} + \dfrac{1}{1680}$, etc., the expression will be

changed into $\dfrac{2}{B\sqrt{n}}$,

He goes on to say that he set himself to determine B but found that the series converged very slowly.

My worthy and learned friend, Mr. James Stirling, who applied himself after me to that enquiry, found that the quantity B did denote the Square-Root of the circumference of a circle whose radius is unity, so that if the circumference be called c the ratio of the middle term to the sum of all the terms will be expressed by $2/\sqrt{nc}$. . . . I also found that the logarithm of the ratio which the middle term of a high power has to any term distant from it by an interval denoted by l would be denoted by a very near approximation (supposing $m = \frac{1}{2}n$) by the quantities

$$(m + l - \tfrac{1}{2}) \log (m + l - 1) + (m - l + \tfrac{1}{2}) \log (m - l + 1)$$
$$- 2m\log m + \log((m + l)/m).$$

Corollary 1. This being admitted, I conclude, that if m or $\frac{1}{2}n$ be a quantity infinitely great, then the logarithm of the ratio, which is a term distant from the middle by the interval l has to the middle term, is $-2l^2/n$.

Corollary 2. The number which answers to the hyperbolic logarithm $-2l^2/n$ being

$$1 - \frac{2l^2}{n} + \frac{4l^4}{n^2} - \frac{8l^6}{6n^3} + \frac{16l^8}{24n^4} - \frac{32l^{10}}{120n^5} + \frac{64l^{12}}{720n^6} \text{ etc.,}$$

it follows that the sum of the terms intercepted between the middle and that whose distance from it is denoted by l will be

$$\frac{2}{\sqrt{nc}} \text{ into } l - \frac{2l^3}{1.3n} + \frac{4l^5}{2.5n^2} - \frac{8l^7}{6.7n^3} + \frac{16l^9}{24.9n^4} - \frac{32l^{11}}{120.11n^5} \text{ etc.}$$

Now let l be supposed $= s\sqrt{n}$, then the said term will be expressed by the series

$$\frac{2}{\sqrt{c}} \text{ into } s - \frac{2a^3}{3} + \frac{4s^5}{2.5} - \frac{8s^7}{6.7} + \frac{16s^9}{24.9} - \frac{32s^{11}}{120.11} \text{ etc.}$$

Moreover, if s be interpreted by $\frac{1}{2}$ then the series will become

$$\frac{2}{\sqrt{c}} \text{ into } \frac{1}{2} - \frac{1}{3.4} + \frac{1}{2.5.8} - \frac{1}{6.7.16} + \frac{1}{24.9.32} - \frac{1}{120.11.64} \text{ etc.}$$

which converges so fast that by the help of no more than seven or eight terms the sum required may be carried to six or seven places of decimals.

There are several other corollaries and lemmas, all bearing on slightly different aspects of computation of terms. Corollary 6 is interesting in that he remarks that if $l = \sqrt{n}$ the series will not converge as fast as it did for $l = \frac{1}{2}\sqrt{n}$ and that it is then necessary to

> make use of the Artifice of Mechanic Quadratures, first invented by Sir Isaac Newton and since presented by Mr. Cotes, Mr. James Stirling, myself and perhaps others.

De Moivre then makes two Remarks, following the great James Bernoulli. In Remark 1 he considers his approximation for increasing n and shows that

> although chance produces Irregularities, still the odds will be infinitely great that . . . those Irregularities will bear no proportion to the recurrency of that order.

In Remark 2 he gives the inverse argument.

> As upon the supposition of a certain determinate law according to which any event is to happen, we demonstrate that the ratio of the happenings will continually approach to that law as the experiments or observations are multiplied : so *conversely*, if from numberless observations we find the ratio of events to converge to a determinate quantity, as to the ratio of P to Q, then we conclude that this ratio expresses the determinate law according to which the event is to happen.

He gives a little more and then says :

> As it is thus demonstrable that there are in the constitution of things certain laws according to which events happen, it is no less evident from observation that those laws serve to wise and beneficent purposes : to preserve the stedfast order of the universe, to propagate the several species of beings and furnish to the sentient kind such degrees of happiness as are suited to their state.
>
> But such laws, as well as the original design and purpose of their establishment, must all be *from without:* the inertia of matter and the nature of all created beings rendering it impossible that anything should modify its own essence or give to itself or to anything else, an original determination or propensity. And hence, if we blind not ourselves with metaphysical dust we shall be led by a short and obvious way, to the acknowledgement of the great MAKER and GOVERNOUR of all, *Himself all-wise, all-powerful and good.*

And here we will leave him—blinded by metaphysical dust, poverty-stricken and with failing eyesight, at the age of 87—the first of the great analytic probabilists.

References

Most of the biographical detail of this chapter I have taken from

H. M. WALKER, *Abraham de Moivre*

in the Journal *Scripta Mathematica* (2), supplemented by the *Éloge* written by Grandjean de Fouchy ; Hœffer has little if anything more· It is doubtful if much can be added to H. M. Walker's account and references unless further documents become available. One has the feeling that there are more to be found. The letters between de Moivre and James Bernoulli are given in

K. WOLLENSCHLÄGER, " Der Mathematische Briefwechsel zwischen Johann I. Bernoulli und Abraham de Moivre," *Verhandlungen der Naturforschenden Gesellschaft in Basel,*

and will be reprinted in the forthcoming volumes of Bernoulli correspondence. The derivation of what is in essence the Poisson limit was pointed out by

E. M. NEWBOLD, " Practical Applications of the Statistics of Repeated Events," *J. Roy. Statist. Soc.*, **90.**

Karl Pearson's note on de Moivre and the Normal Curve will be found in the journal *Biometrika,* **16.**

Appendix 1

Translated by Jean Edmiston

MEMORABLE ARITHMETIC

or

A Brief and Compendious Treatise of Arithmetic, not only for novices, but also for men experienced and well versed in this art, on account of its aiding of memory, truly necessary ; by William Buckley, Regius of Cambridge ; written some time ago, now first published. London, 1567.

The Book to the Reader

Whosoever wishes to memorize arithmetic while the mind is young, read me ; I am small, but I bring great rewards.

Buckley pays enormous dividends in recompense, since he offers in numbers to make numbers memorable.

To the benevolent reader

It is an old argument, that the perception of arithmetic is abstruse and difficult, like the rest of the arts, and that having been grasped, unless use and diligence are employed, it is difficult to retain. Which when W. Buckley—a learned man with the judgment of Matthew—found to be true, he committed the whole of arithmetic to verse. First, in order that his industry might remove the argument which then perplexed him ; next, that he might make the use of arithmetic more familiar to himself, as he did not want to be ignorant in any way ; lastly, that he might promote the cause of letters, whose development he always studied much to preserve with dignity. In which, although he was not pursuing polish and ornament, which the material itself repudiated, yet he truly and earnestly pursued literalness, so that they were able to propagate it in print with great profit, for the great pleasure of readers. For whatever the books of other men, which every now and then deter the studious by their prolixity,

179

are accustomed to offer, is contained in this, briefly and clearly enough ; yet he has considered the memory very successfully, so that written in this way, in harmonious speech, it may be seen not only to aid it, but even to strengthen it. You, O reader, should enjoy this, and do it with humanity, so that, though Buckley has now been dead some years, this child of his brain may become immortal. Also Buckley, in case you did not know it, was a native of Lichfield, employed at Cambridge in King's College. Thence, having run the gambit of scientific and academic honours, he was urgently summoned to the house of important friends with great promises. Here indeed he settled, but in a short time he grew dear to Edward VI, the king of happy memory, and to his noblemen, who called him a wonder of nature on account of his admirable skill in mathematical disciplines ; so that, fate hastening upon him, when dead he left behind great grief for him.

MEMORABLE ARITHMETIC

By G. Buckley of Cambridge

Part 1. The Science of Numbering

Arithmetic teaches the knowledge of numbers. There are seven parts of it : to count correctly, to add, to subtract, to multiply well, to divide, to progress, and to extract roots with the appropriate skill.

Laws

In numbers the first position is that which is furthest to the right ; the next one forsooth is that which is nearer to the left hand. That number is first which is written in the first place. A circle is never placed at the extreme end.

Concerning numeration

Whatever quantity a number is worth, numeration shows, and describes any number whatsoever by its figures. In order that you may make the figures of the numbers and their places correctly, you must learn them. The figures of the numbers are ten, all of which signify something separate, except the last : which is said to be the figure of nothing ; this is called circle by some, by others cypher ; it is made to fill up the space, not to signify anything. These characters, if they are placed in the first position, mean simply themselves ; and placed second, they mean ten, whereas if the third place is allotted to them, the sum extends to a hundred. The fourth place only tells you the thousands, and the fifth contains the fourth ten times, and the sixth exceeds that by the same again. What more? When any place follows, ten is used to multiply the one before.

The manner moreover of writing numbers, and also of describing them

When you are about to write a number, make a beginning on the right, continuing to the left, until you have written them all. Next, set quaternary marks with small points, and however many points there are, they indicate so many thousands to you. The description of numbers in fact is made from the left.

The Division of numbers

A number is triform, a digit, or an articulate, or a compound one ; a digit is always less than ten, as five, two, four, three, six, seven, eight and nine. An articulate is one which makes ten equal parts, such as fifty, sixty, thirty and ten, and the figure of the cypher is prefixed to every such number. A compound number is one which is made up from these, of which kind are all whose last figure is not a cypher.

Rules

Addition, subtraction, multiplication are done from the right. Division and the extraction of roots, from the left.

On Addition

Addition joins many numbers in one. First write the numbers to be added in this order, placing the first figure under the first, next the second under the second, likewise the third under the third, and in this way place the remaining figures, whatever they may be ; a straight line is drawn below under these.

Having done this, collect all the figures which are in the first row. If a digit should result, it is at once written directly beneath. If it is an articulate one, then a cypher is placed below, and the articulate is added to the following row. If a compound one is produced, you write the digit, and the articulate, as before, you join to the following row of numbers.

The order and position which the row follows having been set forth in this way, you must attempt the next one, and you do again as you did in the first place, until, if there are several, you have collected all the rows. The number written below gives you the desired sum. Subtract nine as many times as you can from the numbers you are adding and note the remainders, also subtract nine from the total. If you see the remainder to be equal to the remainder from before, it is an indication that your work is free from error.

On Subtraction

Subtraction takes away equal from equal number, or takes it from a greater one, and makes it less. Write the number from which the subtraction is to be made above, the lower one which is being taken away being written directly beneath, the first under the first, as is done in addition ; the numbers having been arranged, draw a straight line beneath.

Then take the first figure of the first row of the lower from the figure placed above, writing whatever remains below. You must use a similar method for the remaining rows, and, if you have done it right, the number written below will have the total left. If the amount of the number being subtracted was greater than that which was placed above, then an articulate is taken away from the following row and added to the number, by which it makes it enough for the lower number to be subtracted from it : and you should always add one to the following lower number.

Test

Add the number subtracted to the remaining sum. If this shows the original number, the subtraction is correct. If it is less, you will know you have made a mistake.

On Multiplication

To multiply is to derive a number from a number. From the reckoning of this a number is produced, which contains in itself as many times the number which was multiplied as that multiplying contains one. That number is written first which is to be multiplied, and, the multiplier having been placed directly under this, a line is then drawn in the usual way.

Multiply the whole of the multiplicand by the first figure of the multiplying number, writing whatever is produced below, and if you are multiplying by many figures, bring each to the multiplicand, always writing below whatever is produced ; this is written directly under the multiplying number.

And because, however many the numbers you are multiplying may be, there must necessarily be as many numbers produced, therefore join all by addition.

The number written underneath is rightly called the product. For, as you will see, it always produces the answer.

Test

Divide the product by the multiplying number. If you have not erred, it will give you the multiplicand.

On Division

Division shows any number of parts of a number. The number being divided is placed in the uppermost position, and, two straight lines having been drawn under it, the divisor is placed on the left-hand side under these. Then see how many times the divisor is contained in the number placed above, and this number of times should be placed in the space. And afterwards multiply the divisor by the same, and take away the total which is produced by this from the upper number, placing the remainder above, running through the number from which subtraction was made. If it is a matter of many figures being contained in the second, you should move the divisor towards the right by one figure. It is necessary to inquire again how many times the divisor may be in that which is divided, and to place the quotient within the space as before. Thus you destroy the remainder entirely, whatever is left. Nor does this work differ at all, nor vary from the former. But if that which is divided should be less than the lower number, the upper being left untouched, the divisor is advanced and a nought is placed in the middle space, and provided that you do this you will divide the whole sum.

A Short Summary of the whole thing

Divide, multiply, subtract, and carry over the divisor.

Method of writing the remainder

If anything is left after the division has been made, it is customary to write it all above the divisor, and to draw a small line between those numbers, by which the number may be seen to be a fraction, not a whole number.

Test

You should multiply the quotient by the divisor. Add the remainder, if there is any, to the product, and the original number will appear, unless error has laid hold on you.

On Arithmetical progression

If many numbers exceed each other equally, the progression amasses these easily into a fixed sum. Add the first number to the last. You should multiply the sum by the number showing the concatenation (i.e. number of terms in the series). Half the product shows the required sum.

On Geometrical progression

When the given numbers progress by unequal increases, yet there is a single ratio between them, the progression is said to be one of geometry.

You take the last number as multiplier, which shows whence proportion takes its name. You take away the first from the product (of the last and the first), and divide the remainder by a number which should be equal to the number by which the remaining ones were multiplied, minus one. The quotient shows the sum desired.

Test of progressions

Subtract the proposed series from the whole sum. If you have done nothing wrong, you will find nothing is left.

Of the extraction of square roots

To square is to multiply a number once by itself. To find a root is to search for a number which, multiplied by itself, will produce the given sum. If you wish to find the root of any number, write it. Define alternate figures by marking with points, and place two straight lines beneath it. And because division is like the present matter, begin to work from the point towards the left, by seeking under this the digit which, multiplied by itself, is able to produce either the whole or a great part of the number marked with the point. And the digit is written in the middle space under the point, and then multiplied by itself; take the product from the number above, writing the remainder as when you divide. The quotient is doubled, the first figure of this product, if there are two, is placed to the right, under the number which has no dot appearing above it, and the remaining numbers are placed on the left side. Thus a new divisor emerges, of which you must find how many times it is contained in the number placed above ; place the quotient discovered in the space under the following point. Multiply this first by itself, then by the same divisor, (so that) the two products together make one sum, which is subtracted from the upper number, and you should write the rest in the usual way. Again, whatever the double line suggests to you is doubled, and the new divisor will be the doubling : divide this into the number remaining in the upper place, and finish the rest. Square, multiply, subtract this, place the remainder above. Do this until you have run through all the numbers. If it is ever impossible to have the doubling in the remainder, place a nought in the space, and continue the division.

Test

Square the root, add the remainder to the square ; if the number, whatever it may be, comes out the same as at first, it is right ; if not, the work will have to be done again.

Method of collecting fractions from the remainder

One number is added to the double of the root ; then write the remaining number above this product, a small line being added to it, which separates both numbers.

End of the First Part of Memorable Arithmetic

The Second Part of the Memorable Arithmetic of W. Buckley, Regius of Cambridge

Of fractions

Thus far we deal with integer numbers : now it is necessary to describe the parts of them, and the order and position.

Method of writing and expressing fractions

Any part of a number is described by two numbers, which we call denominator and numerator. The latter is always written above, the former below ; with these it is customary to place a little line between them. Let it be done in due form ; it is the custom, the expression for this case.

The denominator shows the parts of the integer ; from these, the numerator shows how many are being taken. When they are the same, that signifies the integer. If the upper one is smaller, the part is less than the integer. If the upper one is larger, it is necessary to add more to the integer.

Concerning fractions of fractions

If a part has particles, form these in the same way before you proceed, and this should be by multiplying the denominators by each other, and the numerators likewise, and their products have that which the former numbers contained.

Method of reducing integers to fractions

You multiply the whole number by the denominator. You must write that number under the product ; if there are any parts, add these to the product, before you attempt to write the denominator under the number.

Method of extracting whole numbers from fractions

Divide the upper number by the denominator. It will show you the number of integers ; write the denominator under the remainder if there is anything left.

Reduction to lowest terms

If it can be done, you may reform two numbers, or you may find the smallest number which divides both, which you do thus : take the smaller from the larger until they are alike, which should be the divisor of both, and the two products show you the lowest terms. In dividing, if it comes to one, do not think it possible to find lesser primes, which we are accustomed to call primes contradictorily.

To find the value of a fraction

Calculate the sum from the parts of integer known, divide the number produced by the lower, and the number of times will show you what the fraction is worth.

Reduction to the same denominator

First multiply the lower numbers by each other, and this will make known the common denominator to you. Multiply the first number by the denominator of the second, and its denominator by the numerator of the second, and the numerators make what should be placed over the denominator, which serves for both.

Rule to be observed in addition and subtraction

When fractions are of different kinds, before you add or subtract you must make them of one kind.

Rule for multiplying and division

If whole numbers are mixed with fractions, you must reduce

them all to parts : only then will you be able to multiply and divide, just as if they were fractions. If integers have to be multiplied or divided by parts, place one beneath the integers, as if it were a part, and finish the rest according to the following rules.

Addition in fractions

The numerators together make one sum, and the common denominator is written beneath. If there are many terms, resolve the first ; the third figure is added to the sum produced.

Subtraction

Take away the lesser from the greater numerator, and the denominator must be written below the remainder. If a part has to be subtracted from integers, resolve one integer into as many parts as the denominator contains : and next take away the given parts.

Multiplication

Whole is reckoned in wholes, and part in parts.

Division

First transpose both (parts of) the divisor, and finish the rest as when you multiply.

Certain fractions having been given, to know which of them is the greater

After the parts have been reduced to the same terms, the greater proportion of the sum of this will be to the lower [*sic*], and accordingly you say it is the greater fraction.

The Extraction of square roots in fractions

As in whole numbers, so you search out roots in fractions. Only if the fractions are square numbers : otherwise you will labour in vain seeking true roots.

To choose roots nearest to the true ones in fractions

Multiply the numerator by the denominator ; the root of the product will be the new numerator ; write the first denominator directly under it.

To execute the same more exactly, both with fractions and integers

In squaring a number, set six cyphers before it ; square the product ; the root is divided by a thousand. The integer shows the quotient, and thus the part will remain exact, as for the true root, nor will the thousandth part be lacking.*

Method of forming numbers by the Golden Rule

In the first place is the thing bought, and second the price, next is placed the number which is in question, which is customarily of the same kind as the first one.

Practice of the rule

You should multiply the third by the middle number ; divide the number produced by the first. Whatever it shows, trust it to show the unknown sum.

Test

The first being multiplied by the fourth, the same is produced as the middle numbers make, multiplied by each other.

The Rule of Three reversed

As above, distribute the numbers after that fashion, Then multiply the first by the middle one ; the divisor will be the third.

Test

The third multiplied by the fourth will produce the same as the first and the second made, multiplied together.

Rule of Combinations

As many as the numbers are which we wish to combine, so many will the figure be, by which the proportion is increased, the first cf which is always multiplied by one. Join all these numbers by addition into a product ; take away as many numbers as the combination consists of ; what remains names a number, whence it will be manifest how many different numbers they can make in all respects, if anyone should wish to multiply the given numbers by each other. If nothing is taken away from the aforesaid sum, some parts will remain, which will number that which is the greatest among all the numbers, proceeding from the multiplication of the numbers by each other.

---

*The Latin text of this paragraph appears to be faulty—*Translator's note.*

Rule of Association

The conditions of all similar things are in the first place ; money follows in the second ; the third order has perfect states only. When numbers are placed in this way, you follow the law of the Golden Rule, which will give you the desired sum by a simple method.

Rule of Error

Take a certain number, continue with it until you come to a selected number. If error has crept in, you should write this at the side, with a sign whether it is more or less. Next, take another, proceed in the same way, noting the error, if there is any, and the sign. Multiply the error placed first by the later one, and the first error multiplies the following one, in the same way, and if the signs are the same, you should take away the lesser product from the greater, also you should take the greater from the lesser error, and you divide the remainder of the products by the number remaining out of the error. Then you will find the desired number in the quotient. You add the sign, if it varies in sign, to the two products, and collect both errors into the same form, which ought to divide the sum of these products. Then the number sought is shown in the quotient.

The Rule of Alligation

Place the numbers to be alligated arranged in a column, writing first the one to which the comparison is to be made. Then join all the lesser ones to this one, writing the difference of the lesser beside the greater, and vice versa, write the difference of the greater beside the lesser. Then collect all the differences into a definite sum, which will be your first number, and whatever difference may be placed in line with it will be your third number. The number of things being mixed is then set in the middle, or anything else which the inquiry yields : the numbers having been placed in this order, the Golden Rule reduces what remains to the desired end.

END

Epitaph on William Buckley, Deceased

Buckley, longing to discover the hidden mysteries of the heavens, of which no one can ever discover enough, yearned to be carried up and to encamp in the sky and to enter the holy threshold of the infinite God. The Omnipotent heard him, nor did he brutally repulse his vows : he carried Buckley away from the earth, drew him up among the stars. Rapt by the splendour of the marvellous place and of God, now he refuses to replace his foot on earth. Nor does he care to paint the true face of the heavens nor its pleasures, nor to reveal the globe with his accustomed art. Verily he is nourished on ambrosia and enjoys blissful peace, nor does he esteem what the world thinks important. The change of your lot is a happy one, Buckley, well suited to the many gifts of your mind.

G.B.T.H.P.

LONDON. From the house of Thomas Marsh. 1567.

Appendix 2

Translated by E. H. Thorne

SOPRA LE SCOPERTE DEI DADI

(Galileo, *Opere*, Firenze, Barbera, **8** (1898), pp. 591-4)

The fact that in a dice-game certain numbers are more advantageous than others has a very obvious reason, i.e. that some are more easily and more frequently made than others, which depends on their being able to be made up with more variety of numbers. Thus a 3 and an 18, which are throws which can only be made in one way with 3 numbers (that is, the latter with 6.6.6 and the former with 1.1.1, and in no other way), are more difficult to make than, e.g. 6 or 7, which can be made up in several ways, that is, a 6 with 1.2.3 and with 2.2.2 and with 1.1.4, and a 7 with 1.1.5, 1.2.4, 1.3.3, and 2.2.3. Nevertheless, although 9 and 12 can be made up in as many ways as 10 and 11, and therefore they should be considered as being of equal utility to these, yet it is known that long observation has made dice-players consider 10 and 11 to be more advantageous than 9 and 12. And it is clear that 9 and 10 can be made up by an equal diversity of numbers (and this is also true of 12 and 11) : since 9 is made up of 1.2.6, 1.3.5, 1.4.4, 2.2.5, 2.3.4, 3.3.3, which are six triple numbers, and 10 of 1.3.6, 1.4.5, 2.2.6, 2.3.5, 2.4.4, 3.3.4, and in no other ways, and these also are six combinations. Now I, to oblige him who has ordered me to produce whatever occurs to me about such a problem, will expound my ideas, in the hope not only of solving this problem but of opening the way to a precise understanding of the reasons for which all the details of the game have been with great care and judgment arranged and adjusted.

And to achieve my end with the greatest clarity of which I am capable, I will begin by considering how, since a die has six faces, and when thrown it can equally well fall on any one of these, only 6 throws can be made with it, each different from all the others. But if together with the first die we throw a second, which also has six faces, we can make 36 throws each different from all the others, since each face of the first die can be combined with each of the second, and in consequence can make 6 different throws, whence it is clear that such combinations are 6 times 6, i.e. 36. And if we add a third die, since each one of its six faces can be combined with each one of the 36 combinations of the other two dice, we shall find that the combinations of three dice are 6 times 36, i.e. 216, each different from the others. But because the numbers in the combinations in three-dice throws are only 16, that is, 3.4.5, etc. up to 18, among which one must divide the said 216 throws, it is necessary that to some of these numbers many throws must belong; and if we can find how many belong to each, we shall have prepared the way to find out what we want to know, and it will be enough to make such an investigation from 3 to 10, because what pertains to one of these numbers, will also pertain to that which is the one immediately greater.

Three special points must be noted for a clear understanding of what follows. The first is that that sum of the points of 3 dice, which is composed of 3 equal numbers, can only be produced by one single throw of the dice: and thus a 3 can only be produced by the three ace-faces, and a 6, if it is to be made up of 3 twos, can only be made by a single throw. Secondly: the sum which is made up of 3 numbers, of which two are the same and the third different, can be produced by three throws: as e.g., a 4 which is made up of a 2 and of two aces, can be produced by three different throws; that is, when the first die shows 2 and the second and third show the ace, or the second die a 2 and the first and third the ace; or the third a 2 and the first and second the ace. And so e.g., an 8, when it is made up of 3.3.2, can be produced also in three ways: i.e. when the first die shows 2 and the others 3 each, or when the second die shows 2 and the first and third 3, or finally when the third shows 2 and the first and second 3. Thirdly : the sum of points which is made up of three different numbers, can be produced in six ways. As for example, an 8 which is made up of 1.3.4, can be made with six different throws: first, when the first die shows 1, the second 3 and the third 4 ; second, when the first die still shows 1, but the second 4 and the third 3; third, when the second die shows 1, and the first 3 and the third 4; fourth, when the second still shows 1, and the first 4 and the third 3; fifth, when the third die shows 1, the first 3, and the second 4; sixth, when the third shows 1, the first 4 and the second 3.

Therefore, we have so far declared these three fundamental points; first, that the triples, that is the sum of three-dice throws, which are made up of three equal numbers, can only be produced in one way; second, that the triples which are made up of two equal numbers and the third different, are produced in three ways; third, that those triples which are made up of three different numbers are produced in six ways. From these fundamental points we can easily deduce in how many ways, or rather in how many different throws, all the numbers of the three dice may be formed, which will easily be understood from the following table:

1 3 6	10		9		8		7		6		5		4		3	
10	631	6	621	6	611	3	511	3	411	3	311	3	211	3	111	1
15	622	3	531	6	521	6	421	6	321	6	221	3				
21	541	6	522	3	431	6	331	3	222	1						
25	532	6	441	3	422	3	322	3								
27	442	3	432	6	332	3										
108	433	3	333	1												
108		27		25		21		15		10		6		3		1
216																

on top of which are noted the points of the throws from 10 down to 3, and beneath these the different triples from which each of these can result; next to which are placed the number of ways in which each triple can be produced, and under these is finally shown the sum of all the possible ways of producing these throws. So, e.g., in the first column we have the sum of points 10, and beneath it 6 triples of numbers with which it can be made up, which are 6.3.1, 6.2.2, 5.4.1, 5.3.2, 4.4.2, 4.3.3. And since the first triple 6.3.1 is made up of three different numbers, it can (as is declared above) be made by six different dice-throws, therefore next to this triple 6.3.1 a 6 is noted : and since the second triple 6.2.2 is made up of two equal numbers and a third which is different, it can only be produced by 3 different throws, and therefore a 3 is noted next to it : the third triple 5.4.1, being made up of three different numbers, can be produced by 6 throws, therefore a 6 is noted next to it : and so on with all the other triples. And finally at the bottom of the little column of numbers of throws these are all added up : there one can see that the sum of points 10 can be made up by 27 different dice-throws, but the sum of points 9 by 25 only, the 8 by 21, the 7 by 15, the 6 by 10, the 5 by 6, the 4 by 3, and finally the 3 by 1 : which all added together amount to 108. And there being a similar number of throws for the higher sums of points, that is, for the points 11, 12, 13, 14, 15, 16, 17, 18, one arrives at the sum of all the possible throws which can be made with the faces of the three dice, which is 216. And from this table anyone who understands the game can very accurately measure all the advantages, however small they may be, of the *zare*,* the *incontri*,* and of any other special rule and term observed in this game.

---

* *Zara* was a game played with 3 dice, the rules of which were possibly those of the game known to us as Hazard. *Incontri* is used here probably as a technical term in *Zara*.

Appendix 3

Translated by Maxine Merrington

MARIN MERSENNE, 1588-1648

The Life of
The Reverend Father Marin Mersenne
Theologian, philosopher and mathematician, member of the
Order of Minim Fathers
by
Brother Hilarion de Coste of the same Order
printed
in Paris by Sebastian and Gabriel Cramoisy in
1649
by ecclesiastical authority with a dedication to
Louis de Valois, Comte d'Alais.

Licence from the Reverend Father General
Thomas Munos and Spinossa.

The world knows that the Province of Mayne has always been the birthplace of great men, whom learning and courage have made praiseworthy. In the past there were the Cardinals De la Forest, Philastre, du Bellay and Cointereau or Cointerel; Messieurs Guillaume, Seigneur de Langeay, and Martin, Prince d'Yvetot of the House of Bellay, Monsieur de St. François, Master of the Requests, later Bishop of Bayeux, Geofroy Boussard, Chancellor of the University of Paris, Pierre de Ronsard, Jean and Jacques Pelletier, Pierre Belon, Robert Garnier, Felix de la Mote le Vayer, Abel Foulon, Sieur Denisot, Germain Pilon, and in our time Monsieur Coeffeteau, Bishop of Marseille, and a great many others. These are the distinguished men who bear witness to this truth.

The Reverend Father Mersenne was born in this same Province in the borough of Oysé on the 8th of September in 1588, a day celebrated in the Church for the birth of the Virgin Mary, Mother of God, and for the destruction of Jerusalem, which was captured and destroyed by the Emperor Titus, Vespasian's son, as predicted by the Saviour of the world forty years previously. This day is also remarkable for being the birthday of many men famous for their piety, valour and knowledge.

The same day, he was baptised by the Priest, Sieur Pierre Basairdy, through the solicitude of his father and mother, Julien Mersenne and Jeanne Moulière, both pious and excellent people. Sanson Ory and René Blanchar were his godfathers, Marie Mersenne, his paternal aunt, was his godmother, and he was named Marin.

If anyone reproaches me for commenting on these small matters, I would remind him that Plutarch was greatly displeased by the omission of details equally trifling. For he complains of those who did not give in writing the names of the mothers of Nicias, Demosthenes, Formion, Thrasybulus and of Theramenes, notable contemporaries of Socrates ; and on the other hand he was delighted with Plato and Antisthenes, because the first gave the name of Alcibiades' tutor and the second did not disdain naming his nurse. Neglect of the ancient writers has been so great that it has caused a dispute between seven towns which all claim to be the birthplace of Homer; each one claims the honour of being the nurse of the greatest poet of Greece. This has always made me think that it is a grave fault of those who are concerned in writing the lives of famous men ; they leave out small particulars, which though not considered of much importance at the time, would be greatly appreciated in another century.

Just as fountain-makers take it to be a good omen when they see vapour coming out of the earth in the morning, because it is one of the signs which make them hope they will find a good spring ; so in the same way, those who have the best knowledge of our souls, rejoice in noting at a tender age, a passionate desire to learn and a rapturous ardour for knowledge and for virtue, because from that they can conjecture almost certainly the good quality of our minds and the excellence which we must one day achieve.

He, of whose life I am writing, appears to have had a good nature from his earliest years ; he had an ardent inclination towards piety and a noble passion for all kinds of curious and agreeable things : for no sooner could he talk than he only spoke of good things : scarcely could he walk but he wished to go to school. In fact he had an aversion for all other forms of exercise except prayer and study. These two occupations never irked him and the older he grew, the more he discovered the delights of knowledge and religious study ; so much so that one had to use force if one wished him to leave these happy pursuits.

His parents, seeing his inclination to devotion and study, sent him to Le Mans, where he never failed to satisfy the demands of his Masters and where he soon showed, by little victories gained over his companions, that he would one day triumph in the great world of knowledge.

At that time King Henry the Great gave the Royal palace of La Flèche to the Jesuit Fathers in order to establish a College of their Order. No sooner had Marin Mersenne heard this news than he begged his parents to send him there. He studied with these learned men with great facility, not only literature, which because of its sweetness is called a Humanity, but also Logic, Physics, Metaphysics, Mathematics and some works on Theology, on all of which he thrived happily. This made him a favourite with the Fathers Chastelier, De la Tour, Jean Phelipeaux and others.

After he left the College of La Flèche, he came to Paris to continue his studies in that famous University and in the Royal College heard the illustrious Professors Marius Ambosius, George Criton and Theodore Marsile, and in the Sorbonne (where reside the strength and support of the Faith) the three celebrated doctors André du Val, Philippe de Gamaches and Nicolas Ysambert, whose names will be immortal amongst the pious and the learned. Under these great men, he took his course in Theology which he always honoured as the queen of all knowledge, the rest being no more than her servants. Thus he devoted the best part of his life to this holy exercise, having never let a day pass without reading the Bible and some Greek or Latin Father.

It was by means of this holy occupation, and through the good example set by the Minim Fathers of the Convent at Plessis, near Tours (through which he happened to pass on his way to Paris from his native district), that he resolved to join this Order.

He applied for the habit at the Convent of Paris, near the Place Royale, from the Rev. Father Olivier Chaillou, who was Vicar-General at the time. This good Father saw that he was received at the Convent of Notre-Dame de Toutes Graces, also called Nigeon, near Paris, by the Reverend Father Pierre Hébert, who was then the Provincial of the Province of France, a man whose memory is blessed among our people, as much for his piety and for his humility as for the exemplary way in which he governed the Order of which he was the thirty-second General.

Having therefore received the habit of the Order from the hands of the Reverend Father Hébert in the Convent of Nigeon on the 17th of July in the year 1611, feast of the incomparable St. Alexis, and after spending two and a half months there, he was sent for the remaining ten months of his year of Probation to the Convent of St. Pierre de Fublines, near Meaux and the Royal palace of Monceaux, where he made his Profession on the 17th July, 1612, at the age of 24, at the hands of the Venerable Father Nicolas Guériteau, Corrector of this Convent, founded by Monsieur Pierre Poussemie, Canon and Precentor of the Church of St. Estienne of Meaux.

He spent his noviceship most devoutly at the Convents of Nigeon and Fublines, edifying by his virtue and humble scholarship all the men in these two Monasteries, to whom he set a good example in humility, penitence, obedience and charity. This is why he was permitted, by common consent, to take his religious vows, which he had faithfully kept with God, having led a life on earth worthy of Heaven, so poor, so chaste and so pure that he triumphed over those passions which usually triumph over most men, and had sacredly preserved that foremost freedom with which all men are born.

After his religious Profession, the simplicity of his manner, his modest ambition and his love of books and learning enabled him to live contentedly in the Order of the Minims, for innocence reigned in his soul ; in the Cloister, he sought only the acquisition of knowledge and virtue ; the desire to learn and practise the good, and the conversation of wise and pious men were his only pursuits and delights.

Two and a half months after his Profession, he went to live in the Convent of Paris, where he took Orders as Sub-Deacon, Deacon and Priest at the hands of Monseigneur Henry de Gondy, Bishop of Paris, who has since become Cardinal of Raiz; he celebrated his first Mass on the 28th of October in 1613, the Feast of the Apostles St. Simon and St. Jude.

Being a Priest, he learnt to perfection the Holy language which was taught him by the Reverend Father Jean Bruno, the Scot, who is said to have been made a Doctor of Theology at the University of Alcala de Henarez and the University of Avignon before entering the Order of Minims and who since then had established the Order in Flanders or the Netherlands with the Reverend Father Jean Sauvage, the celebrated Preacher of the same Order.

The Reverend Father Jean Prieur, being elected Provincial of the Province of France at Michaelmas in 1614, gave Father Mersenne an order under Holy obedience to stay at the Convent of St. Francis of Paula, which the late Duc de Mantouë, de Montferrat and de Nivernois had founded near the town of Nevers in order to teach Philosophy. And in fact Mersenne taught philosophy there during the years 1615, 1616 and 1617 and taught Theology during 1618. But he was obliged to stop these activities as he was elected Corrector of this same Convent. He governed with all the virtues proper to a Superior of a Religious House.

When he had finished his Correctoriate towards the end of 1619, he received an order under Holy obedience from the Reverend Father Hébert (who was for the second time Provincial of the Province of France) to live as a Conventual at the Convent of the Annunciation and of St. Francis of Paula near the Place Royale ; where he had no sooner arrived than he planned to work on the Holy Scriptures and wrote the first volume of his *Commentaires sur la Genèse*, which saw the light of day in 1623. He dedicated it to Monseigneur Jean François de Gondy, First Archbishop of Paris. He produced at the same time *Des Remarques sur les Problèmes de George Venitien.*

The same year, he presented to the public two small books of devotions in French, namely, *L'Analyse de la Vie Spirituelle* and *L'Usage de la Raison.*

Furthermore, seeing that impiety was growing steadily in that unhappy age and that God was greatly dishonoured by certain young Libertines, he was inspired to refute their detestable maxims in French, as he had already done in Latin in his commentary on Genesis. That was why he published a book divided into two parts and volumes entitled :

L'impiété des Déistes, des Athées, et des plus subtils libertins de ce temps, combattue et renversée de point en point par raisons tirées de la Philosophie et de la Théologie.

He also gave to the public his book *De la Vérité des Sciences,* in which he refuted the opinions of the Sceptics and Pyrrhonists ; and also two small volumes in Latin for Mathematicians called *De l'Abrégé ou Inventaire de la Mathématique* and another book in French called *De l'Harmonie Universelle.*

Next he wrote several other books in the same language called *Les Questions Inouïes* ; *Les Questions Harmoniques* ; *Les Questions Théologiques, Physiques, Morales et Mathématiques* ; *Les Mécaniques de Galilée* ; and *Les Préludes de l'Harmonie.*

He wrote *Douze Livres de l'Harmonie* in Latin, which he revised and augmented in a second edition a few months before his death.

But as he loved his country and honoured his nation greatly, he transcribed this book into French in two large volumes in folio under the title *L'Harmonie Universelle, contenant la Théorie et la Pratique de la Musique.*

In the first Volume he described the nature of sound, rhythm, consonance, dissonance, style, modes, composition, the voice, songs and all kinds of harmonic instruments with their notations.

The second Volume contained the practice of consonances

and dissonances ; figured counterpoint ; a method of teaching and learning singing ; the embellishment of airs ; accentual music ; rhythms ; prosody ; French metrics ; methods for singing the Odes of Horace and Pindar ; the use of harmony ; and several new observations both physical and mathematical.

Three Volumes in quarto written in Latin, of which the first contained the following Treatises, entitled

1. *Des Mesures, des Poids, et des Monnoyes des Hébreux, des Grecs et des Romains, réduites à la valeur de celles de France.*
2. *Des Phénomènes ou secrets naturels qui se font par les mouvements et les impressions de l'eau et de l'air.*
3. *Le moyen de naviger et de cheminer dessus et au dessous des eaux, avec un Traité de la Pierre d'Ayman.*
4. *De la Musique speculative et Pratique.*
5. *Un Traité des Mécaniques selon la Théorie et la Pratique.*
6. He explains the trajectories of bullets, arrows, javelins and similar bodies projected by force from longbows and crossbows.

The second Volume consisted of *Un Abrégé de la Géométrie Universelle et des Mathématiques mixtes* which gave :

Firstly, the fifteen books of the Elements of Euclid, with three others by Monsieur François de Foix de Candale, Bishop of Aire, Commander of the two Orders of the King, the Euclid of his time.

2. Twenty-seven books on the Geometry of Pierre de la Ramée, called Ramus.
3. The works of Archimedes, that is two books on the sphere, the cylinder, the dimensions of the circle, the conic and spherical forms, etc.
4. The supplement of Archimedes.
5. Three books on the spherics of Theodosius, three also on Menelaus, and three on Maurolic and on Antoli of the *Sphère avec Théodose* about the many habitats of men who live on the earth. The Phenomena of Euclid and Cosmography. Four books of Apollonius' conic sections. Two books of Selenus on the section of the cylinder. Four books on the conic sections of Monsieur Mydorge. Eight books summarizing the Collections of Pappus, which give Euclid's suppositions. The section of angles of Monsieur Viète and several other Treatises. Two

books on Mechanics, in which are found the works of Commendius and of Luc. Valerius ; and on the centre of gravity of solid bodies etc. Seven books on Optics, where he explains catoptrics, dioptrics, parallaxes and different aspects of refraction.

In the third Volume is found *Les Nouvelles Observations Physicomathématiques avec Aristarche Samien de la Constitution du Monde.*

We must not omit here that our Reverend Father Mersenne took the trouble to revise the Latin and French book *Thaumaturgue Optique* by the Reverend Father Jean François Niceron the Parisian, a Religious of our Order ; after writing his book this Religious had died at the Convent of Aix-en-Provence on the 22nd of September, 1646, aged only thirty-three, to the great sorrow of the scholars and intellectuals who knew him and who loved him for his great knowledge of Theology, Philosophy and Mathematics and for his other excellent qualities.

While working on this book and at the same time on a second Volume of *Commentaires sur la Genèse, et sur S. Mathieu,* and while making constant experiments on the vacuum, he fell ill on the 27th July, 1648, with an abscess which was thought at first to be a false pleurisy. A few days later, seeing that the illness on his side did not lessen but became worse from day to day, he prepared himself to leave this terrestrial life for the eternal and blessed one, since death which appears frightful to most men seemed to him full of enchantment and beauty. He faced the end of his life with all the tenderness of his heart, having purified it by a scrupulous General Confession of his whole life, which he made to me on the 5th of August, Feast of our Lady of the Snows ; thus he fortified himself by frequent reception of Holy Communion, by the Holy Viaticum and by Extreme Unction which he demanded with insistence and which he received with incredible zeal and fervour. So that having armed himself with these divine weapons for the battle between the flesh and the spirit and having shed all human affections in order to clothe himself with Jesus Christ alone, he resigned himself to this fearful moment as a perfect Christian and a true Religious. The Venerable Father Jean Auvry, Corrector, and all the brethren of this Convent of St. Francis of Paula near the Place Royale, who had looked after him for the thirty-seven days of his illness and who saw him die, wonder yet at the great strength of his character. After having said, during the last days of his illness, what his intentions were about his books which were in the press, and having asked the Father Superior to sequester all the forbidden books that were in his room, his unfettered soul thought only of opening the way to Heaven.

Thus lived, and thus died the Reverend Father Marin Mersenne, Member of the Order of Minims of St. Francis of Paula on the 1st of September at three o'clock in the afternoon in the year 1648, having lived sixty years all but one week. He had been a Religious for thirty-seven years, during which he had spent his time either in praying to God or studying or conferring, as much in conversation as in writing, with many able men in all professions, who respected him greatly not only for his knowledge (for he did not ignore anything which could make a man wise) but also because of his sweetness, his humility and all his other excellent qualities which made him the admiration of all those who had the good fortune to know him either by his discussions or by his writings or by the journeys he made in Germany, Flanders and Holland in 1630, in France in 1639 and in Italy and France during the years 1644, 1645 and 1646. For he made friends with the most distinguished and the most celebrated people of the countries in which he travelled.

He was universally mourned by those who had known him, both great and small. I cannot describe the tenderness of heart he bestowed on all who spoke to him. His discourse was never sad, but it was imbued with a certain ingenuousness and a sweetness so engaging that he seemed to hold a gentle power over men's hearts. In fact everyone loved his conversation above all things.

Sixtin Amama, Professor of Grammar at Franeker in Friesland, and Robert Fludd, English physician from the University of Oxford, wrote books against Father Mersenne : but the first, recognising his frankness and sincerity, later made friends with him, as one may see from the pleasant and worthy letters he often wrote to him. The other, having abused both his person and his books with insults such as might be expected from a man without Religion, had to his great displeasure seen many learned men take sides with Father Mersenne against him ; amongst whom were the Reverend Father François de la Nouë, Parisian, Theologian of our Order of Minims (now Assistant to the Most Reverend Father Thomas Munoz and Spinossa, Corrector General of the same Order) who wrote under the name Sieur Flaminius ; also, the Reverend Father Jean Durel of Forez, Theologian of the same Order, under the name Eusebe de St. Just ; and Monsieur Gassendi, Provost of the Church of Digne in Provence, who refuted by solid arguments the insults, impertinences and false opinions of this enraged and melancholy man.

These two writers have acquired no glory from the books they wrote against Father Mersenne, for instead of being hurt by the jealous taunts aimed at his virtue and knowledge, he caused the self-same arrows to fall back on their heads by the sincerity of his actions, by the probity of his life and by the strength of his doctrine.

Many excellent men (other than the three I have named) have spoken magnificently of Father Mersenne or have praised his works in their books, such as Claude Robert, Canon and Vicar of Châlon-sur-Saône, who in his *Gaule Chrétienne* wrote : " Marin Mersenne from Le Mans deserves to be numbered amongst the most distinguished of the Minim Fathers for his infinite piety and learning."

The twin brothers, Messieurs de Sainte Marthe, worthy Historiographers to the King, wrote of Mersenne in the second edition of Robert's *Gaule Chrétienne* which they had enlarged and in which can be found the Catalogue of Archbishops, Bishops and Abbots of France.

Father Jean Philipeaux of the Order of Jesuits wrote of him in his *Commentaires sur Osée.*

Father George Fournier, Jesuit, in his *Hydrographie.*

Dom Pierre de St. Romvald of the Order of Feuillant Fathers, in the third volume of his *Trésor Chronologique et Historique.*

Father Louis Jacob de St. Charles, of the Order of Carmelite Fathers, in his *Traité des Bibliothèques*.

Michel Florent Langrenus, Mathematician and Cosmographer to the King of Spain, in his *Selenographie* or description of the moon.

Jean Hévélius, magistrate for Danzig in Poland, in his fine, learned and inquiring *Selenographie*.

Bonaventure and Abraham Elzevirs, in the *Préface des Oeuvres Mathématiques de François Viète*, *Poitevin*, Conseiller du Roi and Master of the Requests at the Palace, printed through the good offices of François de Schooten, Professor of Mathematics at Leyden University in Holland.

The Abbot Dom Jean Caramuel Lobkowitz, Religious of the Order of Cistercians, Doctor of Theology in the University of Louvain, in divers treatises on Theology and Mathematics.

Father Luc de Montoya, Religious of our Order of Minims, in his *Préface sur les Métaphores du Livre de la Genèse*.

Father Claude Rangueil of Crépy-en-Valois, Theologian of the same Order, in his commentaries on the Book of Kings.

And also Father Simon Martin, Religious of the same Order, in his Eulogy of Mary, sister of Moses and Aaron.

Jaques d'Auzoles, Sieur de La Peyre, in his *Sainte Chronologie* and his *Mercure Charitable* and also in other books.

René Des Cartes, French gentleman, in his *Réponse aux septièmes Questions* [*sic*].

The Reverend Father Jaques Bolduc, Theologian in the Order of Capuchin Fathers, in his *Commentaires sur Job*.

Christofle Scheinerus, commonly called Scheiner, Jesuit, in his book entitled *La Rose des Ursins* honours Father Mersenne's Commentaries on Genesis most highly, as may be seen on p. 735 where he is mentioned together with Jean Baptiste Follengius and Pierre Hurtado de Mendoza, both Jesuits.

Jean Berovicius or van Beverwich, in the problem he propounded by letter, namely, whether our life can possibly be prolonged or shortened, or whether it is necessarily of limited duration, addressed among others a letter to Father Mersenne in which he called him a most distinguished Philosopher and he also gave the answer he had sent him.

Pierre Meusnier, Doctor of Medicine, addressed an Epistle to Father Mersenne at the beginning of his lectures on Philosophy, in which he describes him as very devout and very scholarly and concludes with the Father's reply to him.

The Reverend Father Valérien Magni, the Milanese, Theologian and Philosopher of the Capuchin Order (amongst whom his name is famous in Italy and Poland for his piety and learning, which made him beloved by the Grand Prince, the late King of Poland and Sweden, Vladislas IV), addressed and dedicated to Father Mersenne his treatise on *L'Athéisme d'Aristote*, printed in Warsaw and dated 19th of November, 1647.

Most authors who have written on experiments which question whether Nature suffers a vacuum, have quoted Father Mersenne ; amongst others the Reverend Father Estienne Noel, Rector of the Jesuit College of Clermont in Paris, on p. 59 of his book *De la Pesanteur comparée, ou de la Comparaison de la pesanteur de l'air avec la pesanteur du Vif-argent*, and he quotes him on p. 104 in Chapter 6 of his *Des Observations Physicomathématiques*.

Monsieur Hobbes, the Englishman, tutor to the Prince of Wales, in his books on Philosophy and Mathematics.

Monsieur Nicolas du Chesne de Forest in his book on Philosophy.

Monsieur Naudé in his *Addition à l'Histoire du Roi Louis XI* and in his *Advis pour dresser une Bibliothèque*.

Monsieur Petit, Controller of Fortifications, who had had continual correspondence with Mersenne about experiments and curious facts, in his *Discours Chronologique*, *Traité du Vuide* and in several other works.

Leon Allatio, the Greek, in his book entitled *Apes Urbanae*, about the distinguished men who published books and who were in Rome during the years 1630, 1631 and 1632, mentions Father Mersenne on p. 115.

In the book entitled *Réfutation d'un libelle imprimé à Rouen sous le titre de Futilité etc.* there is on p. 22 a quotation from Mersenne's work *Des Instruments de Musique*, side by side with Boethius.

John Selden, the Englishman, praised him highly in several of his works ; he admired the goodness of his nature and his scholastic diligence. Those who have read his book *Marmora Arundeliana* will notice that he quotes Father Mersenne's Commentaries on Genesis three or four times on a single page.

Jean Pellius, or Pele, Professor of Mathematics in the new Academy of Breda, also quotes him on p. 55 in his book entitled *Controversia de vera circuli mensura*.

In a word the most polished and most learned men of Europe respected and honoured him as an Oracle.

Guillaume Colletet, Advocate to the Parlement de Paris and to the Conseil d'Etat and Privé du Roi, extolled the Reverend Father Mersenne's rare sufficiency in several places in his History of the French poets ; but particularly in the life of Jacques Pelletier of Le Mans, a learned doctor, excellent poet and very accomplished in Mathematics :

> " But in the immortal sciences," wrote Colletet, " we have today two men who know precisely all that was known by Eudoxus and Hipparchus, those two famous antagonists, who became both the rivals of Euclid and the legitimate successors of Ptolemy ; I am speaking of the Reverend Father Mersenne, Religious of the Minim Order, and Pierre Gassendi, two intellects who, despite the ignorance of the age, represent us in some measure, these two live monuments who, in spite of the waters of the universal deluge, preserve for the world all the arts and all the sciences, in which they excel, the one vying with the other. Furthermore, I can say with truth, that for their great aptitude, they deserve no less esteem from the French than ancient Berosius of Chaldea deserved from the Athenians, who erected a statue of him in precious metal in their schools, of which even the tongue was made of gold. O happy age, O happy empire, where virtue was so nobly rewarded! Besides Father Mersenne's fine and profound knowledge, I also praise his ardent passion for our French poetry ; he even earnestly urges us to tune our holy songs to David's lyre and to compose a sweet and melodious harmony for our verse. The letters he sent me on this subject are the glorious and visible witness of his generous feelings and of his great affection for the Muses."

Monsieur de la Mothe le Vayer, Historiographer to the King, addressed his *Discours Sceptique sur la Musique* to our Rev. Father Marin Mersenne, as I have observed twice in this discourse.

Firstly :

> As if to please you, my Reverend Father, let us descend from the general consideration to the particular one of Music ; I recall that you have had such exalted thoughts on this matter and that the ancients gave us nothing equal to them, we shall nevertheless find certain things to doubt and to which we can apply our sceptical arguments regarding the uncertainty of what strikes our senses. Your profound reflections on that charming game of Mathematics leave no hope of anything being added in the future, for they surpass by far all that the last centuries have given us ; what can you expect of me and of my way of arguing, which you already know, but doubts and vacillations which possess me and which concern me just as much as the better known axioms and the more arresting maxims of the School? I am very sensible of my temerity in troubling you with so small a thing, but since the authority you hold over me deprives me of the liberty of a denial, I believe the crime of resisting you with ingratitude would be much worse than simply to be too bold in obeying you. One dedicates so many days to small things in your Temples that the good intention and sanctity of the place makes it worthwhile. I hope that the first or second consideration will operate here in the same way.

Secondly :

> You will hear nothing else on this subject from me, my Reverend Father, except what, in my opinion, will suffice to comply Sceptically with my prime purpose ; realising the beautiful and rare manner in which you have treated Music leaves me but one way of saying anything after you. I have not hesitated in making play of the fashions of discoursing or the ways of the Epoch, though I know well that you have never disapproved of them while keeping within the limits of human knowledge and that you have never blamed the Sceptic when he is dutiful to Heaven and subjecting reason to obedience to the Faith, being content to attack the pride of the Dogmatists for the incertitude of their discipline.

Monsieur Gassendi, Provost of the Church of Digne and Tutor to the King, in the fifth book of the life of that illustrious man, Nicolas Claude Fabry, Seigneur de Peiresc, Counsellor to the Parliament of Aix, often spoke with respect of Father

Mersenne, and gave him this fine eulogy when he praised the worthy Senator, and the honour of Provence for the favours he bestowed on men of letters, noting how he lent them books from his library, which was one of the best and most rare not only in France but in the whole of Europe.

First, he sent a volume on the Theory of Music to Monsieur Doni and later the same volume and another in Arabic with figures accurately drawn by Father Marin Mersenne of the Order of Minims, a man filled with great goodness and learning and indefatigable in exerting himself to clarify and bring to the light of day Religious truths and secrets of Nature.

Jean Jaques Bouchard, Parisian, in the funeral oration which he made in Rome on the 21st of December, 1637, in the Academy of Humorists in honour of the same Seigneur de Peiresc, in the presence of the Cardinals François and Antoine Barberini, Bentivoglio, de la Cueva, Bisci, Pamphilio (now Pontiff of Rome and called Innocent X), Pallote, de Brancas, Aldobrandini and Borghese and several learned men, not only of Italy but of the whole Christian world who lived in that city : after praising several friends of this learned Counsellor of the Parliament of Provence, who are famous not only for their knowledge but also for holding high offices of state and of the Royal Court of this Kingdom which they fill so worthily, he spoke these words in honour of our Frenchmen who made literature their profession, and amongst whom he includes Father Mersenne :

I speak of Sirmond, Petau, Morin, Mersenne, Bourdelot and Valois and many other men famous for great learning and for their illustrious writings.

Monsieur Ismael Boulliau, Priest, living with M. de Thou, wrote on p. 269 of his Notes on Theon of Smyrna :

Finally, experience shows that for the organ system the division of tones of equal parts makes a more perfect and sweeter chord, and the division of whole octaves into twelve equal semi-tones makes the consonances of the chords between them more perfect. On this subject, one must read the Rev. Father Marin Mersenne in his Treatise on the Organ in his *Harmonie Universelle.*

The same writer in the Prolegomena of his *L'Astronomie Philolaïque,* where he speaks of Monsieur Viète's *Harmonicon Céleste,* wrote :

> He had written a work entitled *Harmonicon Céleste,* which Monsieur du Puy once lent to Father Mersenne of the Order of Minims, to gratify his curiosity, for he was ever searching for new and rare things.

Gabriel Naudé, Parisian, Prior of Artige, Canon of Verdun and Librarian to the Cardinals de Bagni, Antoine Barberini and Mazarin, gave him this eulogy on p. 265 of the Octavo edition of his *Question du Destin et du Terme dernier de la Vie* :

> I can make this clear with plausible arguments. Marin Mersenne and Pierre Gassendi, men born for the public good and for the advancement of the most noble sciences, have firmly proved with most powerful and newly recognised arguments that what is contained in Astrology is supported by neither reason nor experience.

The same writer in his appreciation of that excellent doctor and mathematician, Hierôme Cardan, the Milanese, puts our Father Mersenne into the second order of great intellects :

> The second order of intellects comprises those who have made great progress in diverse fields of knowledge such as Cicero, Plutarch, Pliny, Vives, Gesner, Bodin, Patrizzi, Mazzono, Leon Allatio, Mersenne, Doni and other like men.

The same writer, in his appreciation of Augustin Niphus of Sessa in the kingdom of Naples, the foremost philosopher of his time, wrote :

> I am not so adverse to our Frenchmen that I would wish to conceal the fact that once upon a time a most famous Philosopher appeared brilliantly amongst them, I mean Jean Crassot, and today again we have a Gassendi, a Mersenne, a Boulliau, a Des Cartes and a Beauregard who are well able to defend the honour

and the glory of the most subtle and the most curious Philosophy and to escort her happily into the Palace of Knowledge.

Jean Cecile Frey, physician and eminent Professor of Philosophy in the University of Paris, in his new and easy methods for studying the divine sciences, the arts and the knowledge of languages, wrote :

> That is to say, for the Arts, you can do no better than learn what our small book brings to light, entitled : *Les Arts et les Sciences ordonnées et définies*, and those interested in Mathematics should not miss anything written by Father Mersenne, the most learned of all the Religious.

The Reverend Father Theophile Reynault, Jesuit, in his *Trinitas Patriarcharum*, namely St. Bruno, founder of the Carthusian Fathers ; St. Francis of Paula, founder of the Minim Fathers, and St. Ignatius Loyola, founder of the Jesuit Fathers, wrote on p. 395 this panegyric of our Religious :

> Marin Mersenne, fount of all knowledge, who has written on a prodigious diversity of material, whom this age looks upon with admiration and whom posterity will respect with astonishment. But we must not offend the modesty of a living person.

He was customarily visited by many Prelates, Princes, Seigneurs, Theologians, Counsellors, Doctors, Mathematicians and excellent Poets, whose names are famous for their learning and for their love of literature.

Among the ecclesiastics I have noted :

Monseigneur le Cardinal François Barberini, who was Legate from Pope Urban VIII in this kingdom to the late King of France and Navarre, Louis XIII, the Just.

François de Harlay, Archbishop of Rouen and Primate of Normandy.

Charles de Monchal, Archbishop of Toulouse.

The late Monsieur Louis Bretel, Archbishop of Aix.

The late Monsieur Gabriel de Laubespine, Bishop of Orleans.

The late M. Jean Jaubert de Barrault, Bishop of Bazas, later Archbishop of Arles.

The late M. Gilles de Souvré, Bishop of Comminges, later Bishop of Auxerre.

M. Jean Plantevit de la Pause, Bishop of Lodève.

M. Estienne Puget, Bishop of Dardanie, at present of Marseille.

The late M. Henry de Sponde, Bishop of Pamiez.

M. Antoine Godeau, Bishop of Grasse and of Vence.

M. Isaac Habert, Bishop of Vabres, and Doctor of the Sorbonne.

M. Louis de Bassompierre, Bishop of Xaintes.

The late M. Scipion d'Aquavive d'Arragon, Duc d'Atrie, Canon of St. Peter's, Rome, and Abbot of St. Arnoul de Metz.

M. André de Saussay, Official and Grand Vicar to Monseigneur the Archbishop of Paris, nominated Bishop of Toul by the King.

The late M. Nicolas Claude Fabry de Peiresc, Abbot of St. Marie de Guitres in Aquitaine, and Conseiller du Roi en sa Cour de Parlement de Provence.

M. de Refuge, Abbot of St. Cybar d'Angoulesme and Conseiller du Roi en sa Cour de Parlement de Paris.

M. César d'Estrée, Abbot of Notre-Dame de Longpont.

M. the Abbot of Chambon of the House of Hay in Brittany, Doctor of Theology in the Faculty of Paris.

and many other members of this Faculty of Theology, namely :

M. Chastelain, Canon of Notre-Dame in Paris.

M. Chapelas, Curé of St. Jaques de la Boucherie.

M. Perreret, grand Master of the Royal College of Navarre.

M. Frizon, Doctor of the same House, Counsellor and Chaplain to the King, who wrote the lives of the most eminent Cardinals of France under the title *Gallia Purpurata* in Latin.

M. Jean de Launoy, Doctor of the same House.

M. Bandel, Doctor of the Sorbonne.

M. Bachelier, Doctor of the same House.

M. Pierre Gassendi, Provost of the Church of Digne, known throughout Europe for his learning.

The late M. Jean de Cordes, Canon of Limoges.

M. Jaques Pradier, Abbot of Notre-Dame la Blanche in the Isle de Noirmoustier.

The late Claude Robert, Canon of Châlon.

M. de Nesmes, Canon and Lecturer on Theology of St. Sauveur d'Aix.

The late M. de Gautier, Prior and Seigneur of La Valète in Provence.

The late M. Simon de Muys, Canon of Soissons and King's Professor in the Holy tongue in the University of Paris.

M. le Jay, Dean of Vézeley and Conseiller d'Etat, who, with a special Dispensation, had had the Bible printed in Hebrew and in other Oriental languages.

The Reverend Father Guillaume Gibieuf, Doctor of the Sorbonne and Priest of the Oratoire.

The Reverend Father Jean Morin of Blois, also Priest to the congregation of the Oratoire of our Lord Jesus Christ.

The Reverend Fathers—

Jaques Sirmond, Confessor to the late King Louis XIII,
Pierre Bourdin,
Jean Phelipeaux,
Denys Petau,
Estienne Noel,
Nicolas Caussin, also Confessor to the late King Louis XIII, and Jean François, all seven Jesuits and famous for their published books.

Father Thomas Campanella, the Calabrian, Theologian of the Dominican Order.

Dom Jean de Vassan of the Order of Feuillant Fathers, said to be of St. Paul.

The late Father Dominique de Jesus of the Order of Barefoot Carmelites.

Father Louis Jacob, said to be of St. Charles, of the Carmelite Order, Counsellor and Chaplain to the King.

The Reverend Father Jaques Bolduc.

The Reverend Father Joseph de Morlaix of the Capuchin Order.

Dom Martin Marrier, Benedictine and Prior of the Monastery of St. Martin des Champs, well known for the books he has given to the public.

Dom Michel Bauldri from Le Mans, also a Religious of the Benedictine Order of the Abbey of Notre-Dame d'Evron and Prior of the Abbey of Lagny, who has written a book on the ceremonies of the church.

Father Artus du Moustier.

Father Leonard Duliris of the Order of Récollets.

M. Fremart, Master of Music at Notre-Dame de Paris, highly esteemed for his compositions.

The late M. Abraham Blondet, Succentor of the same Church.

M. Boulliau, Priest, excellent Theologian, Philosopher and Mathematician.

M. Michel du Chesne, Parisian, Professor in Philosophy in the Palace of Navarre, very accurate in his researches into the secrets of Art and Nature, concerning which he has made many experiments, as much in France as in Italy.

M. Joseph Voisin, Priest, Counsellor and Chaplain to Monseigneur Armand de Bourbon, Prince of Conty, born of a good family of Bordeaux, very knowledgeable in Hebrew and Greek, as he has shown in his book on Hebrew Theology and other Treatises.

M. Rebours, Priest, to whom Mersenne dedicated one of his books.

M. Bonard, Priest, Chaplain to the late Mgr. André Fremiot, former Archbishop of Bourges, a man learned in Medicine and Philosophy, in which he made a variety of experiments.

M. Germain Habert, Abbot of Cerisy, who wrote the life of the late Mgr. the Cardinal de Berulle.

M. the Abbot de Launay of the House of Brissonets.

M. Sublet, Abbot of Vandosme, to whom Mersenne dedicated one of his books on Mathematics.

M. de Longueterre of the House of Perrotins de Daufiné, who wrote several excellent books, amongst others : the life of the late Mgr. François de Sales, Bishop and Prince of Geneva and Founder of the Order of La Visitation de St. Marie, *Les Souspirs de Philothée*, and *L'Esclavage de la Vierge*.

M. Thomas of the ancient and noble race of Trinobantes in England, who wrote several books, amongst others, *Trois Dialogues du Monde*, learned in matter, form and in its implications.

M. Abraham, the Maronite, Doctor in Theology and in Philosophy, interpreter to the King in Syrian and Arabic, and Professor of these languages in the University of Paris.

M. Nicolas Forest du Chesne, Professor in Rhetoric, Mathematics, Philosophy and in Theology, of whom I have already
spoken amongst the authors who wrote of Mersenne's work
with admiration.

M. the Abbot de Burzeis, well known for his learning and his
writings, and an infinity of others with whom I am not
acquainted.

Amongst the laity, I have noted :

Monseigneur Louis Emmanuel de Valois, Comte d'Alais,
Colonel in Chief of the Light Cavalry of France, Governor
for the King in his own province and for the army of Provence,
grandson of King Charles IX of glorious memory, a Prince
who cherished learned men no less than soldiers. Thus he
followed the example of that great and liberal Monarch, his
grandfather and of his ancestors the Kings and Princes of the
Royal House of Valois and Angoulesme.

The late Monseigneur Antoine de Bourbon, Comte de Moret,
natural son of King Henry the Great of immortal memory.

The late Prince Christofle, second son of Antoine, King of
Portugal.

M. le Prince de Guemené of the most illustrious House of Rohan.

M. le Duc de Luynes.

The late Monseigneur Jean de St. Bonnet, Seigneur of Toiras,
Mareschal de France.

M. le Marquis de Rouillac of the illustrious House of Got, special
Ambassador for the King in Portugal.

M. Henry de Beringhen, Chevalier Seigneur d'Armainvilliers et
de Grez, Conseiller du Roi en ses Conseils, and his first
Equerry.

Don Vasco Luis de Gama, Comte de la Vidiguera, Grand Admiral of the East Indies, special Ambassador of Jean IV, King
of Portugal and Algarves, to our most Christian Louis XIV, a
prince endowed with great affection for study and learned
men, the grandson of the great Vasco de Gama, first con-
queror of the East Indies, of which he was Admiral and the
second Viceroy.

M. Leon Bouthilier, Comte de Chavigny and Minister of State.
Sir (Kenelm) Digby, English knight, known throughout Europe
for his excellent qualities.
and many other Lords and noblemen of that Kingdom.

M. le Marquis d'Estampes Valençay, Chevalier, Conseiller du
Roi en ses Conseils, formerly His Majesty's Ambassador to the
Ministers of State of the United Provinces, brother of the late
Cardinal de Valençay and of Monseigneur the Archbishop of
Rheims.

The late M. Charles de Laubespine, Seigneur de Verderonne et
de Stors, also Chevalier et Conseiller du Roi en ses Conseils
and Keeper of the Seals to his Royal Highness.

His brother M. Claude de Laubespine, Baron de Norat, a nobleman who is no less dear to the Muses than he is valiant on
the field of battle, who has on many occasions shown his
affection for him (Mersenne).

M. Balthasar Gobelin, Conseiller du Roi en ses Conseils, formerly
President of his Chambre des Comptes.

M. Michel Larcher, Seigneur de la Fortelle, Conseiller du Roi en
ses Conseils and President of his Court of Accounts.

M. Gilbert Gaumin Lhuillier, Seigneur d'Orgeval.

M. Jean Pierre de Montchal,

M. Henry Louis Habert, Seigneur de Montmor et de la Brosse,
all of whom were Conseillers du Roi and Masters of Requests
at the Royal Palace. Mersenne dedicated his books on
Harmony, written in Latin, to M. de Montmor.

The late M. Jean Jaques Barrillon, Seigneur de Chastillon sur
Marne, Conseiller du Roi en sa Cour de Parlement and President in his Enquestes, to whom he (Mersenne) dedicated *Les
Phénomènes Ballistiques*, in which he explains the paths of
bullets.

M. Jaques Auguste de Thou, Baron de Melay, Conseiller en la
Cour de Parlement and also President in the first Court of
Enquestes.

M. Hierôme Bignon, Conseiller d'Etat and Advocate-General in
the Cour de Parlement.

M. Lesné,

M. du Bouchet, Seigneur de Bournonville,

M. Vaideau, Seigneur de Gramon,
all Conseillers in the same Cour.

The late M. de Broussel, also Conseiller du Roi en sa Cour de Parlement and Commissioner of Requests at the Palace.

The late M. André le Fevre, Sieur d'Amboile, Conseiller du Roi en sa Cour de Parlement and Commissioner of Requests at the Palace, eldest son of M. d'Ormesson, Conseiller du Roi en ses Conseils.

M. Marcel, Seigneur de Bouqueval, Conseiller du Roi en son Grand Conseil, to whom he has dedicated *Les Phénomènes Méchaniques*.

M. Bruslard, Seigneur de Saint Martin, formerly Conseiller au Grand Conseil.

M. de Carcavi, of Lyons, formerly Conseiller du Roi en sa Cour de Parlement at Toulouse, and au Grand Conseil.

M. de Fermat, Conseiller au Parlement at Toulouse.

M. d'Espagnet, Conseiller at the Cour at Bordeaux.

M. le Baron de Rians of the House of Fabry in Provence, Conseiller du Roi en sa Cour de Parlement at Aix, and his late father

M. Palamèdes de Fabry, Seigneur de Valavez, worthy brother of the late M. de Peiresc, of whom I have already spoken.

M. François Lhuillier, Conseiller du Roi at Toul.

M. Nicolas Rigaud, also Conseiller in the same Parlement, whose name is famous for his great knowledge of the sciences and for the Greek language, and for giving us the works of Tertullian and other excellent books; he had lent Mersenne several manuscripts from the King's library, when he was Keeper there, as had also done

MM. du Puy, those two illustrious brothers, the one a Conseiller d'Etat and the other Prior of St. Sauveur (whose names will be immortal and a beacon throughout the ages), had on all occasions shown the affection they felt for Mersenne.

M. Halé, formerly Conseiller du Roi and Dean of the Masters of the Court of Accounts, to whom he had dedicated

Remarques sur les Problèmes de George de Venize, and many other books.

The late M. Bigot, Sieur de Gastines, formerly also Master of Accounts.

The late M. le Baron d'Arsy, to whom he had dedicated one of his books.

M. le Chevalier de Montmaigny.

M. de Beruille de la Pommeraye.

M. Querin le Vignon,
M. Pierre Mersenne,
MM. René Moreau, Charles Guillemeau, King's Physician ; Jaques Cornuti,
all doctors of Medicine in the Faculty of Paris, the last three celebrated for their writings.

MM. de la Brosse, de la Chambre, Savot, Tournere and du Clos, doctors in the Faculty of Montpellier.

The late Sieur Tournere, physician to the late Duchesse d'Orleans, de Chartres and de Montpensier, and to the Duchesse de Guyse and de Joyeuse, and chief physician to the Lord Chancellor, who had left marks of his learning and his eloquence in excellent works.

The late M. Savot, physician to the late M. le Président Jeannin.

These excellent Mathematicians :

M. de Pagan.

M. Bourdin, Seigneur de Villaines.

M. Claude Mydorge, Treasurer General of France at Amiens.

M. Claude Hardy, Conseiller du Roi at the Chastelet de Paris.

M. G. Pers. de Roberval, Royal Professor of Mathematics at the College of Maître Gervais, and in the chair of Ramus in the Royal College of France, whom Mersenne, on his death, charged with the publication of *Les Traités de la Dioptrique et de la Catoptrique*, to be placed after the books on Optics by the late Father Jean François Niceron.

M. le Tenneur, formerly Conseiller in the Cour des Aydes of Guyenne.

M. Jean Baptiste Morin, Physician and tutor to the King in Mathematics.

M. Tevenot, nominated Resident for the King at Gennes.

MM. Pascal, father and son, the father had formerly been President in the Cour des Aydes in Auvergne.

M. de Beaune, Sieur de Gouliou, formerly Conseiller in the Presidial at Blois.

The late M. Boulenger, tutor to the King in Mathematics, and Preceptor to the late Monseigneur Louis de Bourbon, Comte de Soissons, Prince of the Blood and Peer of France.

The late M. Sanclarus, also Professor to the King in Mathematics.

The late M. Cotel, Conseiller du Roi in the Cour des Aydes.

M. Taffin, his secretary.

M. Picques, the elder, Secretary to the King.

M. Picques, the younger, Advocate in the Cour de Parlement.

The late M. Beaugran, Secretary to the King,

and the late M. Donaut,

both excellent Mathematicians.

M. Gaigneres, Secretary to the late Duc de Bellegarde.

M. de Mets, Commissaire des guerres.

M. Antoine le Comte, Conseiller du Roi, Secretary and Controller General of War.

M. Clercelier.

The late M. Paul Yvon, Sieur de la Leu.

The late M. Jean Tileman Stella, native of Sighen in the County of Hesse in Germany, Professor of Mathematics to the King.

M. Desargues, who was particularly interested in lightening the the work of artisans by the subtilty of his inventions, as for example in the cutting of stones and other things.

M. Girard, tutor to the late Monseigneur François de Valois, Comte d'Alais.

The late M. Gilles Magne, tutor to the late Monseigneur Eleonor d'Orléans, Duc de Fronsac.

M. de Lozières de Nemours, tutor to MM. de Gondrin and de Montespan.

M. I. Mittanour, Astronomer to Monseigneur Armand de Bourbon, Prince de Conty and Prince of the Blood.

These illustrious writers on Philosophy, History, Music and Poetry :

M. René des Cartes, French gentleman, son and brother of Conseillers au Parlement of Brittany, who lived for a while in Holland withdrawn from high society in order to study philosophy more easily.

M. Picot, who translated des Cartes' *Principes de la Philosophie* into French.

M. Marandé, Greffier to the Cour des Aydes, well known for the books he has published.

The late M. Jean Bourdelot, who had an excellent knowledge of good books and Oriental languages, and who had given to the public *Les Oeuvres de Lucian*, which he had translated from Greek into Latin.

M. Naudé, renowned for his excellent qualities and for his knowledge, who had been chosen by three Cardinals as Director of their excellent Libraries, and who can be called without flattery, a living Library, because of his great knowledge of both sciences and books.

M. François de la Mothe le Vayer, Counsellor and Historiographer to the King, formerly Conseiller du Roi and Deputy for the Procureur in the Parlement de Paris.

MM. Scévole and Louis de Sainte-Marthe, twin brothers, Advocats en Parlement and King's Historiographers, two eyes and two luminaries of the Royal Genealogy.

Pierre Scévole and Nicolas Charles de Sainte-Marthe, children of Scévole and grandsons of the great Scévole or Gaucher de Sainte-Marthe.

MM. Henry and Adrien Valois, also brothers, and worthy Historiographers to the King.

M. Louis Chantereau le Febure, Conseiller du Roi en ses Conseils.

M. le Chevalier de l'Escale, who has written several books, amongst others the life of Cardinal Gilles Albornos.

The late M. Hugues Grotius, Resident for the Queen of Sweden to our Kings Louis XIII and XIV, whose name is celebrated by his books.

M. Pierre d'Hozier, Sieur de la Garde Chevalier de l'Ordre du Roi, Genealogist and Chief Justice of the French army.

M. Marc de Vulson, Sieur de la Colombière, also Chevalier de

Saint Michel, Steward to the King and Gentleman in waiting to His Majesty, who wrote a book *La Science Héroique* and two volumes entitled *Le vrai Théâtre d'Honneur et de Chevalerie,* or *Le Miroir Héroique de la Noblesse.*

Jaques Mauduit, Keeper of the Deposit of Requests, to whom Mersenne had written a eulogy in his *Harmonie Universelle,* where he remarks that France, even during the life of that excellent musician, will honour him with the title of the Father of Music because he has engendered such magnificent music in this Kingdom by the excellence of several compositions and concerts for the voice and for all kinds of musical instruments ; such as had never before been done so perfectly.

His son, Louis Mauduit, Prior of St. Martin de Brétheucourt near Dourdan, an excellent poet and mathematician.

The late M. Boisset, Superintendent of the King's Music.

M. Pierre du Ryer, Secretary to Monseigneur the Duc de Vandosme, who has translated so elegantly and so faithfully the Histories of Herodotus into our language, most of the works of Cicero, the two volumes of *L'Histoire de la Guerre de Flandre* by the Rev. Father Famian Strada of Rome and a member of the Jesuit Order, and who had published several fine Tragedies.

MM. Frenicle, celebrated for their poetry and for their knowledge of mathematics.

MM. Guillaume Colletet and François Colletet, his son, who are also well known for their verse and for the books they have given to the public.

M. Jean Jules César de Villeneuve, a gentleman who is no less cherished by the Muses than he is for his valour on the battle-field.

The late M. André Jumeau, Prior of Sainte Croix and Tutor to Monseigneur Henry de Bourbon, Bishop of Metz and Marquis de Verneüil.

M. le Maire, excellent musician and mathematician.

M. Petit, Controller of Fortifications, of whom I have already spoken in this book.

The late M. Guillaume Passart, excellent Geometrician.

M. René Trouillard with whom Mersenne had done many experiments.

M. Gabriel Michel, Sieur de la Roche-Maillet, Angevin, Advocat en la Cour de Parlement.

M. François de Colombelle, Sieur de Beruille, and Chevalier de l'Ordre du Roi.

M. Elie Deodati, Advocat en la Cour de Parlement.

The late M. Poisson, Sieur de la Bodinière, poet in both Latin and French, son of Pierre Poisson, Sieur de la Bodinière, Counsellor in the Courts of Justice at Angers, who had written *L'Harmonie Chronologique des Histoires de la quatrième Monarchie selon l'ordre des années, ensemble l'Etat de l'Eglise.*

After Mersenne's death, many letters were found in his cell written to him by :

M. le Cardinal François Barberini, nephew of Pope Urban VIII.

M. Louis de Valois, Comte d'Alais, Governor for the King in his own province and of the army in Provence, and grandson of King Charles IX.

Jean Baptiste Baliani, gentleman, from Genoa.

M. des Noyers, Secretary to Louise-Marie de Mantua, Queen of Poland and of Sweden, from Warsaw.

The Reverend Father Valerien Magni, a learned Milanese Capuchin, who also wrote from Warsaw.

M. Chanut, Conseiller du Roi en ses Conseils, President of the Tresoriers Generaux of France in Auvergne, and Resident for the King in Stockholm to Queen Christina of Sweden, from Gothie and from Wandalie.

M. Constantin Huyghens, Secretary to the Prince of Orange, from The Hague and other places in Holland.

The late M. Jean Charles, Comte de Conopaskij, Abbot of Tinez, from Vachory in Poland.

Jean Hévélius, Sheriff, from the town and republic of Danzig in the same Kingdom.

Laurens Eichstadius, Physician, from the same town of Danzig.

Joh. Mochingerus from the same town.

M. Denis de Salvaing, Seigneur de Boissieu, Conseiller d'Etat and

First President of the Court of Accounts of Daufiné, from Grenoble.

MM. de Ponnat and de Coste, Conseillers en la Cour de Parlement of Daufiné, from Grenoble.

M. Jaques de Valois, the Scot, Tresorier General of France in Daufiné, great Astronomer, from Grenoble.

Le Seigneur Jean-Baptiste Doni, gentleman of Florence, from Florence.

Le Seigneur Torricelli, Professor to the Grand Duke in Mathematics, and disciple of Galileo Galilei, from the same town.

Sir (Kenelm) Digby, Resident to the Pope, for Henrietta Maria of France, Queen of Great Britain.

M. le Chevalier Cassian du Puy, commonly le Cau del Pozzo.

Le Seigneur Angelo Ricci.

M. Luc Holstenius, commonly called Holstein, native of Hamburg in Germany, Canon of St. Pierre.

Father Athanase Kerker also Aleman, of the Company of Jesus of Rome.

The late M. Aubert le Mire, native of Brussels, Doyen of Notre-Dame of Anvers, and Precentor to the late Infanta Isabelle-Claire-Eugénie, from Anvers.

M. Milon, Advocat en la Cour de Parlement from several towns in this Kingdom.

M. the Abbot of Monflaines.

M. Stanyhurst, Irishman, Doctor of Theology.

M. Titelouse, excellent musician, from Rouen.

Father Theophile Reynault, Jesuit, from Avignon and from Rome.

Father Claude Richard, Jesuit, from Madrid.

MM. Deschamps and Brun, from Bergerac.

Father Vatier, Jesuit, from La Flèche.

M. de Vienne, Abbot of St. Martin at Nevers.

M. de Meru,

Father Jean François, Rector of the Jesuit College at Nevers, from the same town.

M. Fermat, Conseiller en la Cour de Parlement at Toulouse, from the same town.

M. d'Espagnet, Conseiller en la Cour de Parlement at Bordeau ⸱ from the same town.

M. du Verdus,

M. Trichet, also from Bordeaux.

M. le Tenneur from Tours and from other towns in this Kingdom.

Nicolas Cabeus, of Ferrara, Jesuit, from Rome.

Father Cavallieri from Bologna.

Father Honorat Fabri, Jesuit, from Lyons.

M. de Neuré from Aix and from Lyons.

MM. de Colombi and Borrilli from Aix.

M. Claude Saumaise from several towns in Holland.

M. René des Cartes from the same country.

M. de Schooten,

I. Golius,

Le Sieur Sorbere, all from Leiden.

M. Langrenius from Brussels.

Sixtin Amama from Franeker in the year 1627.

Gilbert Voetius from Utrecht.

Ch. Ravius, Professor of Hebrew in Utrecht.

Jean-Albert Bannius, excellent musician, from Haarlem.

André Rivet of Poitiers, from The Hague.

M. du Laurens from the same town.

Isaac Beertman, Mathematician, from Amsterdam.

Christofle Sturanus from Bremen.

Alexandre Morus from Geneva.

Jean Buxtorfe from Basel.

Dantius,

Seldenus,

Theodore Haat from London.

Thomas Hobbes, Englishman, Tutor to the Prince of Wales.

Le Chevalier Candysh.

Henry Revery,

Chavenius, the Dane ; and from other foreigners.

Amongst these letters were also found some from the members of our Order, who are famous for their writings such as :

The Reverend Fathers—

Jean François
Estienne Octoul
André Real

from Marseilles, Aix and Avignon in the Province of Provence, and from the Province of Toulouse and from Aquitaine the Reverend Fathers :

J. la Combe,
Pierre d'Aguts from Toulouse,
Jaques Bremant from Carcassone ;

and in the Province of France from :

The late Reverend Father Robert Regnault, who had been Conseiller du Roi en sa Cour des Aydes in Paris before entering the Order of Minims, whom one can call the Founder of the fine Library of the House of the Place-Royale, who wrote from Constantinople, Calais and places in the Levant ;

in the Province of Lyons :

The Reverend Father Gabriel Thibaut, from Chaumont in Auvergne,

The Reverend Father Bannier, from Clermont,

Rev. Father Jean du Rel or du Relle, from Moulins and from Lyons.

But particularly from Rome :

Rev. Father Emanuel Maignan, native of Toulouse, Reader in Theology at the Convent of the Holy Trinity at Monte Pincio and one of the French Minims in Rome, who published an excellent book this year, 1648, called *Des Horloges et des Quadrans Solaires*.

The late Father Jean François Niceron.

Rev. Father François de la Noüe, the representative in France for the most Reverend Fathers Laurens de Spezzane and Thomas Munos and de Spinossa, Generals of our Order, who had written in Latin *La Chronique general de l'Ordre des Minimes*, in which, writing of the authors belonging to our Order, he praised the Rev. Father Marin Mersenne in these terms : " celebrated Theologian, Philosopher, and Mathematician, verily πολύγραφος".

Thus it may truthfully be said that there is nothing he has not written about with enlightenment and knowledge. I can give him no better praise than that given by Cardinal Cesar Baronio, the great chronicler of this age, to the late M. Nicolas le Fèvre, tutor to the late King Louis XIII and to the late Prince: " Learning more eminent and more modest has never been seen." It is difficult to describe the love and enthusiasm which he devoted to anything which could contribute to the advancement of knowledge. The books he published show so great a variety of subjects that one cannot believe that he could learn even a part of it, though his works gainsay one. He stimulated virtuous emulation amongst learned men, for he obliged them to give to the public the truths which they had discovered, also to apply themselves earnestly to search diligently into those things which are most obscure, of which many have been discovered in this century, even perhaps more than will be discovered in another. If he was unable to persuade great geniuses to bring their work to the light of day, he tried to force them into it, by inserting in his own books what he had learnt from them, thus showing them that they could easily undertake what was half done, or at least by this virtuous artifice he would prevent posterity from being deprived of some things that would have otherwise died. He did this in several places in his books, and always acknowledged the authors with frankness and sincerity and only printed their work for their advantage and glory.

He had a great aversion for idleness, and no sooner had he left the company of people who had favoured him with a visit, or had ceased reading books, holy or secular, or finished the psalmody at divine service than he went walking in the gardens with us ; these were the recreations he enjoyed and we ourselves enjoyed, when he discoursed on the flowers, fruits, plants and the least of animals and indeed on all the marvels of the Lord which were in our path. He often sang the first verses of Psalm 22 : Dominus regit me, etc., or some paraphrase in Latin or French verse of this same Psalm, or he sang the last verse of the last Psalm : Omnis spiritus laudet Dominum, or the whole of this Psalm which contains an exhortation to praise the Holiness of God in His Saints and His power in His visible works with all manner of harmonious instruments.

The souls of such people have the same quality as that fountain admired by the great Alexander in Babylon, whose waters immediately became illuminated by the rays of the sun, or as soon as they were shown fire; for these good people are so fine and purified and have a vision so clear and brilliant that they are enflamed by the smallest spark to meditate on the things of Heaven and the love of God.

Translator's note. The original, and sometimes inconsistent, spelling of proper names in this Appendix has not been altered.

Appendix 4

Translated by Maxine Merrington

LETTERS BETWEEN FERMAT AND PASCAL AND CARCAVI, translated from *ŒUVRES DE FERMAT*, vol. ii, edited by P. Tannery and C. Henry (1894), Paris, Gauthier-Villars et fils

FERMAT TO PASCAL

1654*

Sir,

If I try to make a certain score with a single die in eight throws ; and if, after the stakes have been made, we agree that I will not make the first throw ; then, according to my theory, I must take in compensation 1/6th of the total sum, because of that first throw.

Whilst if we agree further that I will not make the second throw, I must, for compensation, get a sixth of the remainder which comes to 5/36th of the total sum.

If, after this, we agree that I will not make the third throw, I must have, for my indemnity, a sixth of the remaining sum which is 25/216th of the total.

And if after that we agree again that I will not make the fourth throw, I must again have a sixth of what is left, which is 125/1296th of the total, and I agree with you that this is the value of the fourth throw, assuming that one has already settled for the previous throws.

But in the last example in your letter (I quote your own words) you suggest that if I undertake to get the six in eight throws and

---

* This letter is undated. It answers a letter from Pascal which has not yet been found.

having thrown three times without getting it, and if my opponent wants me to abandon my fourth throw and is willing to compensate me because I might still be able to get the six, then 125/1296th of our total stakes would be due to me.

This, however, is incorrect according to my theory. For, in this case, since the one who is throwing has gained nothing in the first three throws, the total sum remains in play, and the one who holds the die and who agrees not to make the fourth throw should take 1/6th of the total stakes as his reward.

And if he had made four throws without getting the required number and it is agreed he should not throw a fifth time, he should still have the same amount for his indemnity, that is 1/6th of the total stake. As the whole sum remains in play, it follows not only from theory but it is indeed common sense that each throw must have the same value.

I beg you to let me know whether we agree in principle, as I believe, and only differ in application.

I am, with all my heart, etc.,

FERMAT

PASCAL to FERMAT

Wednesday, 29th July, 1654

Sir,

1. Like you, I am equally impatient, and although I am again ill in bed, I cannot help telling you that yesterday evening I received from M. de Carcavi your letter on the problem of points (" les partis ")*, which I admire more than I can possibly say. I have not the leisure to write at length, but, in a word, you have solved the two problems of points, one with dice† and the

---

* " Les partis " means here the allotting of the total stakes among players according to their relative expectations, in the case where the game is abandoned before the end.

† " Le parti des dés " of which they are talking is simply when a person throwing dice bets that he will get a certain number in a stated number of throws.

other with sets of games (" Parti des parties ")* with perfect justness ; I am entirely satisfied with it for I do not doubt that I was in the wrong, seeing the admirable agreement in which I find myself with you now.

I admire your method for the problem of points even more than that for dice. I have seen several people obtain that for dice, like M. le Chevalier de Méré, who first posed these problems to me, and also M. de Roberval : but M. de Méré never could find the true value for the problem of points nor a method for deriving it, so that I found myself the only one to know this ratio.

2. Your method is very sound and is the one which first came to my mind in this research ; but because the labour of the combination is excessive, I have found a short cut and indeed another method which is much quicker and neater, which I would like to tell you here in a few words : for henceforth I would like to open my heart to you, if I may, as I am so overjoyed with our agreement. I see that truth is the same in Toulouse as in Paris.

Here, more or less, is what I do to show the fair value of each game, when two opponents play, for example, in three games, and each person has staked 32 pistoles.

Let us say that the first man had won twice and the other once ; now they play another game, in which the conditions are that, if the first wins, he takes all the stakes, that is 64 pistoles, if the other wins it, then they have each won two games, and therefore, if they wish to stop playing, they must each take back their own stake, that is, 32 pistoles each.

Then consider, Sir, if the first man wins, he gets 64 pistoles, if he loses he gets 32. Thus if they do not wish to risk this last game, but wish to separate without playing it, the first man must say : " I am certain to get 32 pistoles, even if I lose I still get them ; but as for the the other 32, perhaps I will get them, perhaps you will get them, the chances are equal. Let us then divide these 32 pistoles in half and give one half to me as well as my 32 which are mine for sure." He will then have 48 pistoles and the other 16.

---

* " Le parti des parties " is clearly explained in the letter, paragraphs 2 to 6.

Let us suppose now that the first had won two games and the other had won none, and they begin to play a new game. The conditions of this new game are such that if the first man wins it, he takes all the money, 64 pistoles ; if the other wins it they are in the same position as in the preceding case, when the first man had won *two* games and the other, *one*.

Now, we have already shown in this case, that 48 pistoles are due to the one who has two games : thus if they do not wish to play that new game, he must say : " If I win it I will have all the stakes, that is 64 ; if I lose it, 48 will legitimately be mine ; then give me the 48 which I have in any case, even if I lose, and let us share the other 16 in half, since there is as good a chance for you to win them as for me." Thus he will have 48 and 8, which is 56 pistoles.

Let us suppose, finally, that the first had won one game and the other none. You see, Sir, that if they begin a new game, the conditions of it are such that, if the first man wins it he will have *two* games to *none*, and thus by the preceding case, 56 belong to him ; if he loses it, they each have one game, then 32 pistoles belong to him. So he must say : " If you do not wish to play, give me 32 pistoles which are mine in any case, let us take half each of the remainder taken from 56. From 56 set aside 32, leaving 24, then divide 24 in half, you take 12 and give me 12 which with 32 makes 44."

Now, in this way, you see, by simple subtraction, that for the first game, 12 pistoles of the other man's money are due to him, for the second another 12 ; and for the last, 8.

Now, to make no mystery of it, since you understand it so well, and I only wish to see that I have made no mistake, the value (by which I mean only the value of the opponent's money) of the last game of *two* is double that of the last game of *three* and *four* times the last game of four and eight times the last game of *five*, etc.

3. But the proportion for the first game is not so easy to find : it is as follows, for I do not wish to falsify anything, and here is the method which I have tried out so much, because indeed it pleases me greatly.

Being given as many games as you wish, find the value of the first.

Let the given number of games be, for example, 8. Take the first eight even numbers and the first eight odd numbers thus :

2, 4, 6, 8, 10, 12, 14, 16
and 1, 3, 5, 7, 9, 11, 13, 15.

Multiply the even numbers in the following way : the first by the second, the product by the third, the product by the fourth etc. ; multiply the odd numbers in the same way : the first by the second, the product by the third, etc.

The last product of the even numbers is the denominator and the last product of the odd numbers is the numerator of the fraction which expresses the value of the first one of eight games : if each person stakes the number of pistoles expressed by the product of the even numbers, he would get from the opponent's stake the product of the odd numbers.

This can be shown, but with a great deal of trouble, by combinatorial methods like those you have worked out, and I have not been able to demonstrate it by this other method which I have just explained to you but only by combinations. And here are the propositions which lead to it, which are really arithmetical propositions bearing on combinations, which I find have certain beautiful properties :

4. If from any number of letters, for example, 8 :

$A, B, C, D, E, F, G, H,$

you take all the possible combinations of 4 letters and then all the possible combinations of 5 letters, and then of 6, of 7 and of 8, etc., and thus you take all possible combinations starting from that number which is half the total, I say that if you add together half the combinations of 4 with each of the combinations of the numbers above, the sum will be the nth term of the fourth progression which begins at 2, where n is half the total.

For example, I will put it in Latin, for French is no use for this. [Translated from Latin.] If, of any eight letters taken at random,

$A, B, C, D, E, F, G, H,$

all combinations of 4 are added, then of 5, of 6, etc., until 8 :
I say, if you add half the combinations of 4, that is 35 (half 70)
with all the combinations of 5, namely 56, plus all the combinations of 6, namely 28, plus all the combinations of 7, namely
8, plus all the combinations of 8, namely 1, you get the fourth
number in the fourth progression whose origin is 2. I say the
fourth number because 4 is half of 8.

For the numbers of the fourth progression, whose origin is 2, are
$$2, 8, 32, 128, 512, \text{etc.}$$
of which 2 is the first, 8 is the second, 32 third and 128 fourth,
and this 128 equals

35, half the combinations of 4 letters
+ 56 combinations of 5 letters
+ 28 combinations of 6 letters
+ 8 combinations of 7 letters
+ 1 combination of 8 letters

5. That is the first proposition which is purely arithmetical ;
the other concerns the theory of games and is as follows :

In the first place, it must be said that if one has gained one out
of 5 games, for example, and if one needs 4, the match will be
decided for certain in 8 which is double 4.

The value in terms of the opponent's stakes, of the first game
in a set of 5, is a fraction whose numerator is half the combinations of 4 out of 8 (I take 4 because it is equal to the number
of games required and 8 because it is double 4) and whose
denominator is this same numerator plus all the combinations
of higher numbers.

Thus, if I have won the first game out of 5, 35/128 of my
opponent's stake is due to me : that is to say, if he has staked
128 pistoles, I take 35 and leave him the remainder, 93.

Now this fraction 35/128 is the same as 105/384 which is
made by taking the product of even numbers as denominator and
the product of odd numbers as numerator.

You will undoubtedly understand all this well, if you take a
little trouble : that is why I think it unnecessary to go on with it
any longer.

6. I am sending you, nevertheless, one of my former Tables ; I have not time to copy it out, I will remake it.

You will see there, as always, that the value of the first game is equal to that of the second, which is easily shown by combinations.

You will see, in the same way, that the numbers in the first line are always increasing ; so also are those in the second ; and those in the third.

But those in the fourth line are decreasing, and those in the fifth, etc. This seems odd.

		If each one stakes 256 in					
		6	5	4	3	2	1
		games	games	games	games	games	game
	1st game	63	70	80	96	128	256
From my	2nd game	63	70	80	96	128	
opponent's	3rd game	56	60	64	64		
256 pistoles	4th game	42	40	32			
I get, for the	5th game	24	16				
	6th game	8					

		If each one stakes 256 in					
		6	5	4	3	2	1
		games	games	games	games	games	game
	the first game	63	70	80	96	128	256
From my	the first 2 games	126	140	160	192	256	
opponent's	the first 3 games	182	200	224	256		
256 pistoles	the first 4 games	224	240	256			
I get, for	the first 5 games	248	256				
	the first 6 games	256					

7. I have not time to send you the proof of a difficulty which greatly puzzled M. de Méré, for he is very able, but he is not a geometrician (this, as you know, is a great defect) and he does not even understand that a mathematical line can be divided *ad infinitum* and believes that it is made up of a finite number of points, and I have never been able to rid him of this idea. If you could do that, you would make him perfect.

He told me that he had found a fallacy in the theory of numbers, for this reason :

If one undertakes to get a six with one die, the advantage in getting it in 4 throws is as 671 is to 625.

If one undertakes to throw 2 sixes with two dice, there is a disadvantage in undertaking it in 24 throws.

And nevertheless 24 is to 36 (which is the number of pairings of the faces of two dice) as 4 is to 6 (which is the number of faces of one die).

This is what made him so indignant and which made him say to one and all that the propositions were not consistent and that Arithmetic was self-contradictory : but you will very easily see that what I say is correct, understanding the principles as you do.

I will put everything I have done so far in order when I have finished the geometrical Treatises which I have been working on for some time.

8. I have also done some arithmetic on a subject, of which I beg you to let me know your opinion.

I state the lemma which everyone knows : that the sum of as many numbers as you like of the progression which starts at unity,

$$1, \ 2, \ 3, \ 4,$$

taken twice, is equal to the last one, 4, multiplied by the next largest one, 5 : that is to say that the sum of the numbers contained in A, taken twice, is equal to the product

$$A(A + 1)$$

Now I come to my proposition :
[Translated from the Latin]

The difference of any two adjacent cubes, when unity is subtracted, is six times all the numbers (i.e. the sum of the numbers) contained in the root of the smaller one.

R, S, are two roots differing by one. I say that

$R^3 - S^3 - 1$ is equal to the sum of the numbers contained in S multiplied by six.

Let S be called A, then R is $A + 1$.

Therefore the cube of the root R, or $A + 1$, is

$$A^3 + 3A^2 + 3A + 1^3.$$

But the cube of S, or A, is A^3 and the difference of these is

$$3A^2 + 3A = R^3 - S^3 - 1.$$

But twice the sum of the numbers contained in A, or S, equals, from the lemma,

$$A(A + 1), \text{ that is } A^2 + A.$$

Therefore six times the sum of the numbers contained in A equals

$$3A^2 + 3A.$$

But

$$3A^2 + 3A \text{ equals } R^3 - S^3 - 1,$$

Therefore $R^3 - S^3 - 1$ equals six times the sum of the numbers contained in A, or S. Q.E.D.

No one has made any objection to this, but I have been told that no one has objected for the reason that nowadays everyone is familiar with this method : and as for me I maintain that, without flattering myself, one must admit that this proof is of a good, general kind. I await, nevertheless, your opinion with all deference.

Everything I have shown in Arithmetic is of this nature.

9. Here are two further difficulties :

I have demonstrated a proposition in a plane concerning the cube of a line compared with the cube of another : I maintain that it is pure geometry and I say this in all seriousness.

Similarly I have solved the following problem : " Of four planes, four points and four spheres, any four are given, find one sphere which, touching the given spheres, passes through the given points and leaves on the planes portions of the spheres containing given angles."

And this one : " Of three circles, three points, three lines, any three being given, find a circle which, touching the circles and the points, leaves on the lines an arc containing a given angle."

I have solved these problems in a plane, using only circles and straight lines in the construction ; but in the proof I have used parabolas and hyperbolas instead of solids : I maintain, nevertheless, since the construction is in a plane, my solution is in a plane and must pass muster as such.

This is poor recognition of the honour you do me in suffering my wearisome discourse for so long. I intended to say no more than two words to you and I have not yet told you what is closest to my heart, and that is the more I know you the more I admire and honour you, and if you could see how much that is you would find a place in your affections for him who is, Sir, your etc.

PASCAL

FERMAT to CARCAVI

Sunday, 9th August, 1654

Sir,

1. I was overjoyed to have had the same thoughts as those of M. Pascal, for I greatly admire his genius and I believe him to be capable of solving any problem he attempts. The friendship he offers is so dear to me and so precious that I shall not scruple to take advantage of it in publishing an edition of my Treatises.

If it does not shock you, you could both help in bringing out this edition, and I suggest that you should be the editors: you could clarify or augment what seems too brief and thus relieve me of a care which my work prevents me from taking. I would like this volume to appear without my name even, leaving to you the choice of designation which would indicate the author, whom you could qualify simply as a friend.

2. Here is the course which I have thought out for the second Part which will contain my researches on numbers. It is a work which is still only an idea, and for which I may not have the leisure to put fully on paper ; but I will send a summary to M. Pascal of all my principles and first theorems, in which, I can promise you in advance, he will find everything not only novel and hitherto unknown but also astounding.

If you combine your work with his, everything will succeed and soon be completed, and we will thus be able to publish the first Part which you have in your care.

If M. Pascal approves of my overtures which are based on my

great esteem for his genius, his knowledge and his intellect, I will first begin to inform you of my numerical results. Farewell.

I am, Sir, your very humble and very obedient servant,

FERMAT

Toulouse 9th August, 1654.

PASCAL TO FERMAT

Monday, 24th August, 1654

Sir,

1. I could not give you all my thoughts concerning the problem of points for several players in the last post, and now I feel reluctant to do so for fear that the admirable agreement between us which is so precious to me, might begin to wane, for I am afraid we may have different opinions on this subject. I want to show you all my arguments, and you will do me the favour of correcting me if I am wrong and supporting me if I am right. I ask you this earnestly and sincerely for I will consider myself in the right only when you agree with me.

When there are only *two* players, your combinatorial method is very reliable, but when there are *three*, I think I can prove that it is not applicable, unless you proceed in some other way which I have not understood. But the method I have shown you and which I always use can be applied in all cases and in all types of the problem of points, whereas the combinatorial method (which I use only for particular cases because it is shorter than the general method) is only good for those few occasions and not for others.

I am sure that I will make myself understood but I will have to discourse a little and you must have a little patience.

2. This is your procedure when there are *two* players : If two players, playing several games, find themselves in that position when the first man needs *two* games and the second needs *three*, then to find the fair division of stakes, you say that one must know in how many games the play will be absolutely decided.

It is easy to calculate that this will be in *four* games, from which you conclude that it is necessary to see in how many ways four games can be arranged between two players, and one must see how many combinations would make the first win and how many the second and to share out the stakes in this proportion. I would have found it difficult to understand this if I had not known it myself already ; in fact you had explained it with this idea in mind. Thus, to find out how many combinations of four games there are between two players, one must imagine that they play with a die of two faces (since there are only two players) as in heads and tails, and that they throw four of these dice (because they play four games) ; and now one has to see in how many different ways these dice can turn up. This is easy to calculate, it is sixteen altogether, which is the square of four. Now let us suppose that one of the faces is marked *a*, and is favourable to the first player, and the other *b*, favourable to the second player : thus these four dice can turn up in one of the following sixteen ways :

a a a a	*a a a a*	*b b b b*	*b b b b*
a a a a	*b b b b*	*a a a a*	*b b b b*
a a b b	*a a b b*	*a a b b*	*a a b b*
a b a b	*a b a b*	*a b a b*	*a b a b*
1 1 1 1	1 1 1 2	1 1 1 2	1 2 2 2

and because the first player needs two games, he must win whenever there are two *a*'s : thus there are 11 for him ; and because the second player needs 3, he must win whenever there are three *b*'s : therefore there are 5 for him. Thus the stakes must be shared in the ratio of 11 to 5.

This is your method when there are *two* players ; whereupon you say that if there are more players, it will not be difficult to find the fair division of stakes by the same method.

3. Upon this, Sir, I must tell you that the solution of the problem of points for two players based on combinations is very accurate and true, but if there are more than two players it will not always be correct, and I will tell you the reason for this difference.

I showed your method to our colleagues, upon which M. de Roberval made this objection to it.

What is mistaken is that the problem is worked out on the assumption that *four* games are played ; in view of the fact that when one man wins *two* games or the other wins *three*, there is no need to play *four* games, it could happen that they would play *two* or *three*, or in truth, perhaps *four*.

And thus he could not see why you claim to find the fair division of stakes under the pretence that they play *four* games, seeing that in reality the dice will not be thrown again after one of the players has won, and that at least, if it were not false, it had not been proved, so that he had some suspicion that we had made a fallacious argument.

I answered him that I did not rely too much on the combinatorial method, which in truth is not appropriate here, but on my other general method, in which nothing escapes and which carries its own proof and which gives the result as precisely as the combinatorial method ; moreover I showed him the truth of the fair division of stakes between two players by the combinatorial method in this way.

Is it not true that if two players, finding themselves in the hypothetical position when one needs *two* games and the other *three* and they mutually agree to play *four* games, that is to say the four dice with two faces are thrown all together, is it not true, I say, that if they decided to play four games, the correct division of stakes must follow, as we have said, the number of arrangements favourable to each?

He agreed with this, it can in fact be proved, but he denied that the same thing would apply if the players were not compelled to play *four* games. I then replied thus :

Is it not clear that the same two players, not being compelled to play four games but wishing to cease play as soon as one has gained his score, can without loss or gain still decide to play four games, and that this agreement does not change the position in any way at all? For, if the first man wins the first two games out of *four* and thus has won the set, would he refuse to play another two games, seeing that, if he wins them he cannot win more and if he loses them he cannot win less? because these last two which the opponent might win are not enough for him, since he needs three, thus it is impossible for them both to get the number they want in four games.

Certainly it is easy to see that it is absolutely equal and immaterial to them both whether they let the game take its natural course, which is to cease play when one man has won, or to play the whole four games : therefore since these two procedures are equal and immaterial, the result must be the same in them both. Now, the solution is correct when they are obliged to play four games, as I have proved : therefore it is equally correct in the other case.

That is how I proved it, and you will note that this proof is based on the equality of the two procedures, true and hypothetical, regarding the two players ; one or other of the players will always win in either of these two procedures, and if he wins or loses with one procedure, he will win or lose with the other, and never will both players win their point.

4. Let us follow the same argument for *three* players and let us suppose that the first man needs *one* game, the second needs *two* and the third *two*. To solve the problem of points, following the same combinatorial method, one must first find out in how many games the play will be decided, as we did when there were two players : this will be in *three* games, for they cannot play *three* games without reaching a decision.

It is now necessary to find out in how many ways three games can be played by three players, and how many ways will be favourable to one man, how many to the other, and how many to the last, and to distribute the stakes in this proportion, as was done on the supposition that there were two players.

It is easy to see how many combinations there are altogether : it is the third power of 3, that is to say its cube, 27. For, if one throws *three* dice at once (since three games must be played) which each have *three* faces (since there are three players), one marked *a* favourable to the first player, the other marked *b* favourable to the second, the other *c* favourable to the third, it is obvious that these three dice, thrown together, can turn up in 27 different ways, as follows :

a a a	a a a	a a a	b b b	b b b	b b b	c c c	c c c	c c c
a a a	b b b	c c c	a a a	b b b	c c c	a a a	b b b	c c c
a b c	a b c	a b c	a b c	a b c	a b c	a b c	a b c	a b c

1 1 1	1 1 1	1 1 1	1 1 1	1		1 1 1	1	1
	2			2	2 2 2	2		2
		3				3	3	3 3 3 3

Now the first man needs just *one* game to win : thus all the throws giving one *a* are favourable to him : there are 19.

The second man needs *two* games : thus all the throws giving two *b*'s are his : there are 7.

The third man needs *two* games : thus all throws with two *c*'s are his : there are 7.

If from that one concluded that the stakes must be divided amongst the three men in the proportion 19 : 7 : 7 one would make a gross error, and I cannot believe you would do this : for some of the throws are favourable both to the first and second man, such as *a b b*, for the first player has one *a* which he needs and the second has two *b*'s which he needs ; and in the same way *a c c* is favourable to the first and third man.

Thus we must not add up those throws common to two players as being worth the whole amount to each, but only worth half. For suppose *a c c* were thrown, the first and the third man would have the same right to the stake, each having made his point, thus they would each take half the money ; but if *a a b* were thrown the first man would win alone. The calculations must be made in this way :

There are 13 throws favourable only to the first man and 6 which give him a half share and 8 which give him nothing : thus if the stake is one pistole, there are 6 throws each worth $\frac{1}{2}$ pistole to him and 8 worth nothing.

Thus, to share out the stakes, we must multiply

13 by one pistole which makes ..	13
6 by a half, which makes ..	3
8 by zero, which makes ..	0

Total 27   Total 16

and divide the sum of the products, 16, by the sum of the throws, 27, which gives the fraction 16/27 ; this gives the amount due to the first man, when the stakes are shared out, that is 16 pistoles out of 27.

The shares for the second and the third player will be found in the same way :

There are 4 throws worth 1 pistole to him, multiplied together	4
There are 3 throws worth $\frac{1}{2}$ pistole to him, multiplied together	$1\frac{1}{2}$
And 20 throws worth nothing to him	0
Total 27	Total $5\frac{1}{2}$

Thus, $5\frac{1}{2}$ pistoles out of 27 are due to the second player, and the same amount to the third ; these three totals, $5\frac{1}{2}$, $5\frac{1}{2}$ and 16 added together make 27.

5. That seems to me to be the way one must solve the problem of points following your combinatorial method, unless you proceed in some other way which I do not know. But, if I am not mistaken, this solution is unfair.

The reason for this is that you make a false assumption, which is that *three* games are played invariably, instead of letting the play take its natural course, which is to play only until one man has got the number of games he needs when play ceases.

I am not saying that they never play three games, but it could be that they would play only one or two games and no more would be necessary.

But, one will ask, why cannot the same supposition that was made for two players be allowed in this case? Here is the reason :

In the actual conditions of the game with three players, only one man can win, for play ceases as soon as one man has won. But in the hypothetical conditions, two men can get the number of games they need : that is, when the first man wins the single game he needs and one of the others wins the two games he needs : for still they would have played only three games, whereas when there were only two players, the hypothetical and the real conditions fitted in with the interests of both players; it is this which makes such a difference between the real and hypothetical conditions.

But if the players, finding themselves in the hypothetical situation—that is to say, when the first man needs *one* game, the second *two*, the third *two*—mutually agree to the condition that they will play the three games, and that those who will have won the number they need will take the whole stake if they are the only ones to have achieved their point, or if there are two who have got the required number of games they will share out the stake equally, *in this case only* the solution of the problem of points must be as I have just given it : out of 27 pistoles the first man takes 16, the second $5\frac{1}{2}$, and the third $5\frac{1}{2}$, and given this particular situation this is self-evident.

But if they play simply with the condition not that three games must necessarily be played, but only until one of the players has won the number of games he needs, when the game ends without any other player getting a chance to win, then out of 27, 17 pistoles is due to the first man, 5 to the second and 5 to the third.

This can be found with my general method which also shows that in the preceding situation, 16 is due to the first man, $5\frac{1}{2}$ to the second and $5\frac{1}{2}$ to the third without using combinations at all, for it works every time.

6. These, Sir, are my ideas on this subject about which I know no more than you, I have only thought more about it : and that to you is a small thing since your first thoughts are more penetrating than my prolonged efforts.

I do not allow myself to disclose my reasons for awaiting judgement from you. I think I have made you see that, quite by accident, the combinatorial method is suitable for two players as it also is sometimes for three players, for example when one man needs *one* game, another needs *one* and the third *two*, because in this case, the number of games needed to finish the set is not enough for two of the players to win ; but it is not general and is not universally sound, except for the particular case when one is compelled to play a certain precise number of games.

So that, since you did not know my method when you propounded the problem of points for several players, but only the combinatorial method, I am afraid we shall have different opinions on this matter.

I beg you to let me know in what way you proceed when tackling this problem. I shall receive your answer with respect and pleasure, even if your ideas contradict mine. I am, etc.

PASCAL

FERMAT TO PASCAL*

Saturday, 29th August, 1654

Sir,

1. Our sparring continues, and I am as glad as you are that our ideas correspond so exactly; it seems as if they arrived in the same way. Your recent *Traité du Triangle arithmétique et de son Application* is proof of this: and if my reckoning is correct, your eleventh inference came through the post from Paris to Toulouse while my proposition on the theory of numbers, which in effect is the same thing, was going from Toulouse to Paris.

I am not afraid of failing whilst our thoughts cross in this way. I am sure that the best way of avoiding failure is to collaborate with you. But if I say more, it would seem to be a compliment, and we have banished that enemy of sweet and easy conversation.

It is now my turn to give you some of my numerical inventions, but the end of parliament makes me very busy, and I dare hope that, in your goodness, you will allow me a reasonable and almost essential respite.

2. However, I will answer your question about the three players, playing two games. When the first man has one of the games and the others none, your first solution is correct, and the money must be divided in the proportion of 17, 5 and 5: the reason for this is obvious and always follows the same principle,

---

* Fermat wrote this letter before he had received the preceding one (of 24th August 1654).

the combinations making it at once clear that the first man has 17 equal chances, whilst the other two have but 5.

3. For the rest, there is nothing that I will not, in future, communicate quite frankly to you. Meanwhile, if you find it opportune, consider this proposition :

The squared powers of 2 plus one are always prime numbers.

The square of 2 plus one is 5, which is a prime number.

The square of the square is 16 plus one makes 17, a prime number.

The square of 16 is 256 plus one is 257, a prime number.

The square of 256 is 65536 plus one is 65537, a prime number. And so on to infinity.

I will be answerable for the truth of this property. The proof of it is very troublesome and I admit that I cannot find it completely. If I had been successful, I would not suggest that you look for it.

The proposition is useful for finding numbers which can be factorised, concerning which I have made considerable discoveries. We will talk about it another time.

I am, Sir, your, etc.

FERMAT

FERMAT TO PASCAL*

Friday, 25th September, 1654

Sir,

1. Do not fear that our agreement is ending, you yourself have strengthened it in trying to destroy it, and it seems to me that in answering M. de Roberval for yourself you have also replied for me.

I take the example of three players, the first man needing one game and each of the others two, which is the case you have objected to.

I find only that there are 17 combinations for the first man

---

* Answering Pascal's letter of 24th August, 1654.

and 5 for each of the other two : for, when you say that the combination *a c c* is favourable to the first man and to the third, it appears that you forgot that everything happening after one of the players has won is worth nothing. Now since this combination makes the first man win the first game, of what use is it that the third man wins the next two games, because even if he won thirty games it would be superfluous?

The consequence, as you so well remarked, of this fiction of lengthening the match to a particular number of games is that it serves only to simplify the rules and (in my opinion) to make all the chances equal or, to state it more intelligibly, to reduce all the fractions to the same denominator.

So that you shall have no more doubts, if instead of *three* games in the same case, you lengthen the pretended match to *four*, there would be not just 27 combinations but 81, and we have to see how many combinations would give one game to the first man before either of the other two players would get two games, and how many would give two games to each of the other men before the first man would get one. You will find that there are 51 combinations favourable to the first man, and 15 to each of the other two, which comes to the same thing as before.

And if you take 5 games or any other number you please you will always get the same proportions 17, 5, 5.

And thus I am right in saying that the combination *a c c* is favourable only to the first man and not to the third, and *c c a* is favourable only to the third man and not to the first, and therefore my combinatorial rule is the same for three players as for two and, in general, for any number of players.

2. You must have already been able to see from my last letter that I have no doubt at all about the correct solution of the problem with three players, for I sent you the three decisive numbers, 17, 5, 5. But since M. de Roberval would perhaps more easily understand a solution without artifice, here is one for this example : it may also sometimes lead to short cuts in other cases :

The first man can win, either in a single game, or in two or in three.

If he wins in a single game, he must, with one die of three faces, win with the first throw. A single die has three possibilities, this player has a chance of 1/3 of winning, when only one game is played.

If two games are played, he can win in two ways, either when the second player wins the first game and he wins the second or when the third player wins the first game and he wins the second. Now, two dice have 9 possibilities : thus the first man has a chance of 2/9 of winning when they play two games.

If three games are played, he can only win in two ways, either when the second man wins the first game, the third the second and he the third, or when the third man wins the first game, the second wins the second and he wins the third ; for, if the second or third player were to win the first two games, he would have won the match and not the first player. Now, three dice have 27 possibilities : thus the first player has a chance of 2/27 of winning when they play three games.

The sum of the chances that the first player will win is therefore 1/3, 2/9 and 2/27 which makes 17/27.

And this rule is sound and applicable to all cases, so that without recourse to any artifice, the actual combinations in each number of games give the solution and show what I said in the first place, that the extension to a particular number of games is nothing but a reduction of the several fractions to a common denominator. There in a few words is the whole mystery, which puts us on good terms again since we both only seek accuracy and truth.

3. I hope to send you, by Martinmas, a summary of anything worthwhile that I have done on numbers. You will allow me to be concise, for in this way I shall be understood by you who can comprehend the whole from half a word.

What you will find most worthwhile concerns the proposition that every number is composed of one, of two or of three triangles : of one, of two, of three or of four squares ; of one, of two, of three, of four or of five pentagons ; of one, of two, of three, of four, of five or of six hexagons and so on to infinity.

To arrive at this, we must prove that every prime number

which exceeds unity by a multiple of four is composed of two squares, such as 5, 13, 17, 29, 37, etc.

Given a prime number of this kind, such as 53, find by a general rule, the two squares of which it is composed.

Every prime number which exceeds unity by a multiple of 3 is composed of a square and three times another square, such as 7, 13, 19, 31, 37, etc.

Every prime number which exceeds 1 or 3 by a multiple of 8 is composed of a square and twice another square, such as 11, 17, 19, 41, 43, etc.

There is no triangle of numbers whose area is equal to the square of a number.

This would follow from the discovery of many propositions which Bachet admitted he did not know and which are missing from Diophantus.

I am sure that as soon as you know my method of proof for this kind of proposition, you will find it excellent and it will lead to many new discoveries : for it must be, as you know, that *multi pertranseant ut augeatur scientia.*

If I have enough time, later on we will discuss magic numbers, and I will recall my former ideas on this subject. I am, with all my heart, Sir, your, etc.

FERMAT

I wish M. de Carcavi as good health as I have and I think of him always. I write to you from the country and that is why my answers may be delayed during the vacations.

PASCAL to FERMAT

Tuesday, 27th October, 1654

Sir,

Your last letter pleased me well. I admire your method for the problem of dividing the stakes all the more so because I understand it perfectly ; it is yours entirely and has nothing in common with mine ; it reaches the same conclusion very simply. Thus our mutual understanding is restored.

But, Sir, even if I have agreed with you in this, look elsewhere for someone who would follow your numerical discoveries, the explanations of which you have so kindly sent me. I must confess that these are far beyond my comprehension; I can but admire them and very humbly beg you to take the first opportunity of completing them. All our colleagues saw them last Saturday and valued them very highly : it is not easy to wait for such elegant and desirable things.

Please think about them, and be assured that I remain, etc.

PASCAL

FERMAT TO PASCAL

Sunday, 25th July, 1660

Sir,

As soon as I discovered that we were nearer to one another than we had ever been before, I could not resist making plans for renewing our friendship and I asked M. de Carcavi to be mediator : in a word I would like to embrace you and to talk to you for a few days ; but as my health is not any better than yours, I very much hope that you will do me the favour of coming half way to meet me and that you will oblige me by suggesting a place between Clermont and Toulouse, where I would go without fail towards the end of September or the beginning of October.

If you do not agree to this arrangement, you will run the risk of seeing me at your house and of thus having two ill people there at once. I await your news with impatience and am, with all my heart,

Yours ever,

FERMAT

PASCAL TO FERMAT

Tuesday, 10th August, 1660

Sir,

You are the most gallant man in the world and assuredly I am the one who can best recognise your qualities and very much admire them, especially when they are combined with your own singular abilities. Because of this I feel I must show my appreciation of the offer you have made me, whatever difficulty I still have in reading and writing, but the honour you do me is so dear to me that I cannot hasten too much in answering your letter.

I will tell you then, Sir, that if I were in good health, I would have flown to Toulouse and I would not allow a man such as you to take one step for a man such as myself. I will tell you also that, even if you were the best Geometrician in the whole of Europe, it would not be that quality which would attract me to you, but it is your great liveliness and integrity in conversation that would bring me to see you.

For, to talk frankly with you about Geometry, is to me the very best intellectual exercise : but at the same time I recognise it to be so useless that I can find little difference between a man who is nothing else but a geometrician and a clever craftsman. Although I call it the best craft in the world, it is after all only a craft, and I have often said it is fine to try one's hand at it but not to devote all one's powers to it.

In other words, I would not take two steps for Geometry and I feel certain you are very much of the same mind. But as well as all this, my studies have taken me so far from this way of thinking, that I can scarcely remember that there is such a thing as Geometry. I began it, a year or two ago, for a particular reason ; having satisfied this, it is quite possible that I shall never think about it again.

Besides, my health is not yet very good, for I am so weak that I cannot walk without a stick nor ride a horse, I can only manage three or four leagues in a carriage. It was in this way that I took twenty-two days in coming here from Paris. The doctors recommended me to take the waters at Bourbon during the month of September, and two months ago I promised, if I can manage it, to go from there through Poitou by river to Saumur to stay until Christmas with M. le duc de Roannes, governor of Poitou, who has feelings for me that I do not deserve. But, since I go through Orleans on my way to Saumur by river and if my health prevents me from going further. I shall go from there to Paris.

There, Sir, is the present state of my life, which I felt obliged to describe to you so as to convince you of the impossibility of my being able to receive the honour you have so kindly offered me. I hope, with all my heart, that one day I shall be able to acknowledge it to you or to your children, to whom I am always devoted, having a special regard for those who bear the name of the foremost man in the world.

I am, etc.

PASCAL

De Bienassis, 10th August, 1660

Appendix 5

FROM *THE DOCTRINE OF CHANCES*, pp. 243-254, by A. de Moivre, 3rd edition, London, 1756.

A Method of approximating the Sum of the Terms of the Binomial $(a + b)^n$ expanded into a Series, from whence are deduced some practical Rules to estimate the Degree of Assent which is to be given to Experiments.

Altho' the Solution of Problems of Chance often requires that several Terms of the Binomial $(a + b)^n$ be added together, nevertheless in very high Powers the thing appears so laborious and of so great difficulty, that few people have undertaken that Task ; for besides *James* and *Nicolas Bernoulli*, two great Mathematicians, I know of no body that has attempted it ; in which, tho' they have shewn very great skill, and have the praise which is due to their Industry, yet some things were farther required ; for what they have done is not so much an Approximation as the determining very wide limits, within which they demonstrated that the Sum of the Terms was contained. Now the Method which they have followed has been briefly described in my *Miscellanea Analytica*, which the Reader may consult if he pleases, unless they rather chuse, which perhaps would be the best, to consult what they themselves have writ upon that subject : for my part, what made me apply myself to that Inquiry was not out of opinion that I should excel others, in which however I might have been forgiven ; but what I did was in compliance to the desire of a very worthy Gentleman, and good Mathematician, who encouraged me to it : I now add some new thoughts to the former ; but in order to make their connexion the clearer, it is necessary for me to resume some few things that have been delivered by me a pretty while ago.

I. It is now a dozen years or more since I had found what follows : If the Binomial $1 + 1$ be raised to a very high Power denoted by n, the ratio which the middle Term has to the Sum of all the Terms, that is, to 2^n, may be expressed by the Fraction

$$\frac{2A \times (n - 1)^n}{n^n \times \sqrt{n-1}} ;$$

wherein A represents the number of which the Hyperbolic Logarithm is

$$\frac{1}{12} - \frac{1}{360} + \frac{1}{1260} - \frac{1}{1680}, \text{ etc.}$$

But because the Quantity $\dfrac{(n-1)^n}{n^n}$ or $\left(1 - \dfrac{1}{n}\right)^n$ is very nearly

given when n is a high Power, which is not difficult to prove, it follows that, in an infinite Power, that Quantity will be absolutely given, and represent the number of which the Hyperbolic Logarithm is $- 1$; from whence it follows, that if B denotes the Number of which the Hyperbolic Logarithm is

$$- 1 + \frac{1}{12} - \frac{1}{360} + \frac{1}{1260} - \frac{1}{1680}, \text{ etc.}$$

the Expression above-written will become $\dfrac{2B}{\sqrt{n-1}}$ or barely $\dfrac{2B}{\sqrt{n}}$:

and that therefore if we change the Signs of that Series, and now suppose that B represents the Number of which the Hyperbolic Logarithm is

$$1 - \frac{1}{12} + \frac{1}{360} - \frac{1}{1260} + \frac{1}{1680}, \text{ etc.}$$

that Expression will be changed into $\dfrac{2}{B\sqrt{n}}$.

When I first began that inquiry, I contented myself to determine at large the value of B, which was done by the addition of some Terms of the above-written Series : but as I perceived that it converged but slowly, and seeing at the same time that what I had

done answered my purpose tolerably well, I desisted from proceeding farther till my worthy and learned Friend Mr. *James
Stirling*, who had applied himself after me to that inquiry, found
that the Quantity B did denote the Square-root of the Circumference of a Circle whose Radius is Unity, so that if that Circumference be called c, the Ratio of the middle Term to the Sum of
all the Terms will be expressed by $\dfrac{2}{\sqrt{nc}}$.

But altho' it be not necessary to know what relation the number B may have to the Circumference of the Circle, provided its
value be attained, either by pursuing the Logarithmic Series
before mentioned, or any other way ; yet I own with pleasure that
this discovery, besides that it has saved trouble, has spread a
singular Elegancy on the Solution.

II. I also found that the Logarithm of the Ratio which the
middle Term of a high Power has to any Term distant from it by
an Interval denoted l, would be denoted by a very near approximation, (supposing $m = \tfrac{1}{2}n$) by the Quantities
$$(m + l - \tfrac{1}{2}) \times \log (m + l - 1) + (m - l + \tfrac{1}{2}) \times \log$$
$$(m - l + 1) - 2m \times \log m + \log \frac{m + l}{m}.$$

COROLLARY 1

This being admitted, I conclude, that if m or $\tfrac{1}{2}n$ be a Quantity
infinitely great, then the Logarithm of the Ratio, which a Term
distant from the middle by the Interval l, has to the middle Term,
is $-\dfrac{2ll}{n}$.

COROLLARY 2

The Number, which answers to the Hyperbolic Logarithm
$-\dfrac{2ll}{n}$, being
$$1 - \frac{2ll}{n} + \frac{4l^4}{2nn} - \frac{8l^6}{6n^3} + \frac{16l^8}{24n^4} - \frac{32l^{10}}{120n^5} + \frac{64l^{12}}{720n^6}, \text{ etc.}$$

it follows, that the Sum of the Terms intercepted between the Middle, and that whose distance from it is denoted by l, will be

$$\frac{2}{\sqrt{nc}}\text{ into } l - \frac{2l^3}{1\times 3n} + \frac{4l^5}{2\times 5nn} - \frac{8l^7}{6\times 7n^3} + \frac{16l^9}{24\times 9n^4} - \frac{32l^{11}}{120\times 11n^5},$$

etc.

Let now l be supposed $= s\sqrt{n}$, then the said Sum will be expressed by the Series

$$\frac{2}{\sqrt{c}}\text{ into } s - \frac{2s^3}{3} + \frac{4s^5}{2\times 5} - \frac{8s^7}{6\times 7} + \frac{16s^9}{24\times 9} - \frac{32s^{11}}{120\times 11},\text{ etc.}$$

Moreover if s be interpreted by $\frac{1}{2}$, then the Series will become

$$\frac{2}{\sqrt{c}}\text{ into }\frac{1}{2} - \frac{1}{3\times 4} + \frac{1}{2\times 5\times 8} - \frac{1}{6\times 7\times 10} +$$

$$+ \frac{1}{24\times 9\times 32} - \frac{1}{120\times 11\times 64},\text{ etc.}$$

which converges so fast, that by the help of no more than seven or eight Terms, the Sum required may be **carried to six or seven** places of Decimals : Now that Sum will be found to be 0·427812, independently from the common Multiplicator $\frac{2}{\sqrt{c}}$, and therefore to the Tabular Logarithm of 0·427812, which is 9·6312529, adding the Logarithm of $\frac{2}{\sqrt{c}}$, viz. 9·9019400, the Sum will be 19·5331929, to which answers the number 0·341344.

LEMMA

If an Event be so dependent on Chance, as that the Probabilities of its happening or failing be equal, and that a certain given number n of Experiments be taken to observe how often it happens and fails, and also that l be another given number, less than $\frac{1}{2}n$, then the Probability of its neither happening more frequently than $\frac{1}{2}n + l$ times, nor more rarely than $\frac{1}{2}n - l$ times, may be found as follows.

Let L and L be two Terms equally distant on both sides of the middle Term of the Binomial $(1 + 1)^n$ expanded, by an Interval equal to l; let also s be the Sum of the Terms included between L and L together with the Extreams, then the Probability required will be rightly expressed by the Fraction $\dfrac{s}{2^n}$; which being founded on the common Principles of the Doctrine of Chances, requires no Demonstration in this place.

COROLLARY 3

And therefore, if it was possible to take an infinite number of Experiments, the Probability that an Event which has an equal number of Chances to happen or fail, shall neither appear more frequently than $\frac{1}{2}n + \frac{1}{2}\sqrt{n}$ times, not more rarely then $\frac{1}{2}n - \frac{1}{2}\sqrt{n}$ times, will be expressed by the double Sum of the number exhibited in the second Corollary, that is, by 0·682688, and consequently the Probability of the contrary, which is that of happening more frequently or more rarely than in the proportion above assigned, will be 0·317312, those two Probabilities together compleating Unity, which is the measure of Certainty : Now the Ratio of those Probabilities is in small Terms 28 to 13 very near.

COROLLARY 4

But altho' the taking an infinite number of Experiments be not practicable, yet the preceding Conclusions may very well be applied to finite numbers, provided they be great : for Instance, if 3600 Experiments be taken, make $n = 3600$, hence $\frac{1}{2}n$ will be $= 1800$, and $\frac{1}{2}\sqrt{n} = 30$, then the Probability of the Event's neither appearing oftner than 1830 times, nor more rarely than 1770, will be 0·682688.

COROLLARY 5

And therefore we may lay this down for a fundamental Maxim, that in high Powers, the Ratio, which the Sum of the Terms included between two Extremes distant on both sides from the middle Term by an Interval equal to $\frac{1}{2}\sqrt{n}$, bears to the Sum of all the Terms, will be rightly expressed by the Decimal 0·682688, that is 28/41 nearly.

Still, it is not to be imagined that there is any necessity that the number n should be immensely great ; for supposing it not to reach beyond the 900th Power, nay not even beyond the 100th, the Rule here given will be tolerably accurate, which I have had confirmed by Trials.

But it is worth while to observe, that such a small part as is $\frac{1}{2}\sqrt{n}$ in respect to n, and so much the less in respect to n as n increases, does very soon give the Probability 28/41 or the Odds of 28 to 13 : from whence we may naturally be led to enquire, what are the Bounds within which the proportion of Equality is contained? I answer, that these Bounds will be set at such a distance from the middle Term, as will be expressed by $\frac{1}{4}\sqrt{2n}$ very near ; so in the Case above mentioned, wherein n was supposed $= 3600$, $\frac{1}{4}\sqrt{2n}$ will be about 21·2 nearly, which in respect to 3600, is not above 1/169th part : so that it is an equal Chance nearly, or rather something more, that in 3600 Experiments, in each of which an Event may as well happen as fail, the Excess of the happenings or failings above 1800 times will be no more than about 21.

COROLLARY 6

If l be interpreted by $\sqrt{n}$, the Series will not converge so fast as it did in the former Case when l was interpreted by $\frac{1}{2}\sqrt{n}$, for here no less than 12 or 13 Terms of the Series will afford a tolerable approximation, and it would still require more Terms, according as l bears a greater proportion to $\sqrt{n}$: for which reason I make use in this Case of the Artifice of Mechanic Quadratures, first invented by Sir *Isaac Newton*, and since prosecuted by Mr. *Cotes*, Mr. *James Stirling*, myself and perhaps others ; it consists in determining the Area of a curve nearly, from knowing a certain number of its Ordinates A, B, C, D, E, F, etc. placed at equal Intervals, the more Ordinates there are, the more exact will the Quadrature be ; but here I confine myself to four, as being sufficient for my purpose : let us therefore suppose that the four Ordinates are

260　　　　　APPENDICES

A, B, C, D, and that the Distance between the first and last is denoted by *l,* then the Area contained between the first and the

last will be $\dfrac{1 \times (A + D) + 3 \times (B + C)}{8} \times l$: now let us take

the Distances $0\sqrt{n}$, $\frac{1}{6}\sqrt{n}$, $\frac{2}{6}\sqrt{n}$, $\frac{3}{6}\sqrt{n}$, $\frac{4}{6}\sqrt{n}$, $\frac{5}{6}\sqrt{n}$, $\frac{6}{6}\sqrt{n}$, of which every one exceeds the preceding by $\frac{1}{6}\sqrt{n}$, and of which the last is $\sqrt{n}$: of these let us take the four last, *viz.* $\frac{3}{6}\sqrt{n}$, $\frac{4}{6}\sqrt{n}$, $\frac{5}{6}\sqrt{n}$, $\frac{6}{6}\sqrt{n}$, then taking their Squares, doubling each of them, dividing them all by *n,* and prefixing to them all the Sign $-$, we shall have $-\frac{1}{2}$, $-\frac{8}{9}$, $-\frac{25}{18}$, $-\frac{2}{1}$, which must be looked upon as Hyperbolic Logarithms, of which consequently the corresponding numbers, *viz.* 0·60653, 0·41111, 0·24935, 0·13534 will stand for the four Ordinates *A, B, C, D.* Now having interpreted *l* by $\frac{1}{2}\sqrt{n}$, the Area will be found to be $= 0·170203 \times \sqrt{n}$, the double of

which being multiplied by $\dfrac{2}{\sqrt{nc}}$, the product will be 0·27160;

let therefore this be added to the Area found before, that is, to 0·682688, and the Sum 0·95428 will shew what, after a number of Trials denoted by *n,* the Probability will be of the Event's neither happening oftner than $\frac{1}{2}n + \sqrt{n}$ times, nor more rarely than $\frac{1}{2}n - \sqrt{n}$, and therefore the Probability of the contrary will be 0·04572 : which shews that the Odds of the Event's neither happening oftner nor more rarely than within the Limits assigned are 21 to 1 nearly.

And by the same way of reasoning, it will be found that the Probability of the Event's neither appearing oftner than $\frac{1}{2}n + \frac{3}{2}\sqrt{n}$ nor more rarely than $\frac{1}{2}n - \frac{3}{2}\sqrt{n}$ will be 0·99874, which will make it that the Odds in this Case will be 369 to 1 nearly.

To apply this to particular Examples, it will be necessary to estimate the frequency of an Event's happening or failing by the Square-root of the number which denotes how many Experiments have been, or are designed to be taken : and this Square-root, according as it has been already hinted at in the fourth Corollary, will be as it were the *Modulus* by which we are to regulate our Estimation ; and therefore suppose the number of Experiments to be taken is 3600, and that it were required to assign the Probability of the Event's neither happening oftner than 2850 times, nor more rarely than 1750, which two numbers may be varied at pleasure, provided they be equally distant from the middle Sum 1800, then make the half difference between the two numbers 1850 and 1750, that is, in this Case, $50 = s\sqrt{n}$; now having supposed $3600 = n$, then $\sqrt{n}$ will be $= 60$, which will make it that 50 will be $= 60s$, and consequently $s = \frac{50}{60} = \frac{5}{6}$; and therefore if we take the proportion, which in an infinite power, the double Sum of the Terms corresponding to the Interval $\frac{5}{6}\sqrt{n}$, bears to the Sum of all the Terms, we shall have the Probability required exceeding near.

LEMMA 2

In any Power $(a + b)^n$ expanded, the greatest Term is that in which the Indices of the Powers of a and b have the same proportion to one another as the Quantities themselves a and b ; thus taking the 10th Power of $a + b$, which is $a^{10} + 10a^9b + 45a^8b^2 + 120a^7b^3 + 210a^6b^4 + 252a^5b^5 + 210a^4b^6 + 120a^3b^7 + 45a^2b^8 + 10ab^9 + b^{10}$; and supposing that the proportion of a to b is as 3 to 2, then the Term $210a^6b^4$ will be the greatest, by reason that the Indices of the Powers of a and b, which are in that Term, are in the proportion of 3 to 2 ; but supposing the proportion of a to b had been as 4 to 1, then the Term $45a^8b^2$ had been the greatest.

LEMMA 3

If an Event so depends on Chance, as that the Probabilities of its happening or failing be in any assigned proportion, such as may be supposed of a to b, and a certain number of Experiments be designed to be taken, in order to observe how often the Event will happen or fail ; then the Probability that it shall neither happen more frequently than so many times as are denoted by

$$\frac{an}{a + b} + 1$$

nor more rarely than so many times as are denoted by

$$\frac{an}{a + b} - 1 ,$$

will be found as follows :

Let L and R be equally distant by the Interval l from the greatest Term ; let also S be the Sum of the Terms included between L and R, together with those Extreams, then the Probability required will be rightly expressed by

$$\frac{S}{(a + b)^n} .$$

COROLLARY 8

The Ratio which, in an infinite Power denoted by n, the greatest Term bears to the Sum of all the rest, will be rightly expressed by the Fraction $\frac{a + b}{\sqrt{abnc}}$, wherein c denotes, as before, the Circumference of a Circle for a Radius equal to Unity.

COROLLARY 9

If, in an infinite Power, any Term be distant from the Greatest by the Interval l, then the Hyperbolic Logarithm of the Ratio which that Term bears to the Greatest will be expressed by the Fraction

$$= \frac{(a + b)^2}{2abn} \times ll ;$$

provided the Ratio of l to n be not a finite Ratio, but such a one as may be conceived between any given number p and $\sqrt{n}$, so that l be expressible by $p\sqrt{n}$, in which Case the two Terms L and R will be equal.

COROLLARY 10

If the Probabilities of happening and failing be in any given Ratio of inequality, the Problems relating to the Sum of the Terms of the Binomial $(a + b)^n$ will be solved with the same facility as those in which the Probabilities of happening and failing are in a Ratio of Equality.

REMARK I

From what has been said, it follows, that Chance very little
disturbs the Events which in their natural Institution were
designed to happen or fail, according to some determinate Law ;
for if in order to help our conception, we imagine a round piece
of Metal, with two polished opposite faces, differing in nothing
but their colour, whereof one may be supposed to be white, and
the other black ; it is plain that we may say, that this piece may
with equal facility exhibit a white or black face, and we may even
suppose that it was framed with that particular view of shewing
sometimes one face, sometimes the other, and that consequently
if it be tossed up Chance shall decide the appearance ; But we
have seen in our LXXIId Problem, that altho' Chance may produce an inequality of appearance, and still a greater inequality
according to the length of time in which it may exert itself, yet
the appearances, either one way or the other, will perpetually
tend to a proportion of Equality : But besides, we have seen in the
Present Problem, that in a great number of Experiments, such as
3600, it would be the Odds of above 2 to 1, that one of the Faces,
suppose the white, shall not appear more frequently than 1830
times, nor more rarely than 1770, or in other Terms, that it
shall not be above or under the perfect Equality by more than
1/120 part of the whole number of appearances ; and by the
same Rule, that if the number of Trials had been 14,400 instead
of 3600, then still it would be above the Odds of 2 to 1, that the
appearances either one way or other would not deviate from
perfect Equality by more than 1/260 part of the whole : and in
1,000,000 Trials it would be the Odds of above 2 to 1, that the
deviation from perfect Equality would not be more than by
1/2000 part of the whole. But the Odds would increase at a
prodigious rate, if instead of taking such narrow limits on both
sides the Term of Equality, as are represented by $\frac{1}{2}\sqrt{n}$, we double
those Limits or triple them : for in the first Case the Odds would
become 21 to 1, and in the second 369 to 1, and still be vastly
greater if we were to quadruple them, and at last be infinitely
great ; and yet whether we double, triple or quadruple them,
etc. the Extension of those Limits will bear but an inconsiderable

proportion to the whole, and none at all, if the whole be infinite ; of which the reason will easily be perceived by Mathematicians, who know, that the Square-root of any Power bears so much a less proportion to that Power, as the Index of it is great.

What we have said is also applicable to a Ratio of Inequality, as appears from our 9th Corollary. And thus in all Cases it will be found, that *altho' Chance produces Irregularities, still the Odds will be infinitely great, that in process of Time, those Irregularities will bear no proportion to the recurrency of that Order which naturally results from* ORIGINAL DESIGN.

REMARK II

As, upon the Supposition of a certain determinate Law according to which any Event is to happen, we demonstrate that the Ratio of Happenings will continually approach to that Law, as the Experiments or Observations are multiplied : so, *conversely*, if from numberless Observations we find the Ratio of the Events to converge to a determinate quantity, as to the Ratio of P to Q; then we conclude that this Ratio expresses the determinate Law according to which the Event is to happen.

For let that Law be expressed not by the Ratio, $P : Q$, but by some other, as $R : S$; then would the Ratio of the Events converge to this last, not to the former : which contradicts our *Hypothesis*. And the like, or greater, Absurdity follows, if we should suppose the Event not to happen according to any Law, but in a manner altogether desultory and uncertain ; for then the Events would converge to no fixt Ratio at all.

Again, as it is thus demonstrable that there are, in the constitution of things, certain Laws according to which Events happen, it is no less evident from Observation, that those Laws serve to wise, useful and beneficent purposes ; to preserve the stedfast Order of the Universe, to propagate the several Species of Beings, and furnish to the sentient Kind such degrees of happiness as are suited to their State.

But such Laws, as well as the original Design and Purpose of their Establishment, must all be *from without* : the *Inertia* of matter, and the nature of all created Beings, rendering it impossible that any thing should modify its own essence, or give to itself, or to any thing else, an original determination or propensity. And hence, if we blind not ourselves with metaphysical dust, we shall be led, by a short and obvious way, to the acknowledgement of the great MAKER and GOVERNOUR of all ; *Himself all-wise, all-powerful* and *good*.

Mr. *Nicholas Bernoulli*, a very learned and good Man, by not connecting the latter part of our reasoning with the first, was led to discard and even to vilify this Argument from *final Causes*, so much insisted on by our best Writers ; particularly in the Instance of the nearly equal numbers of *male* and *female* Births, adduced by that excellent Person the late Dr. *Arbuthnot*, in *Phil. Trans*. No. 328.

Mr. Bernoulli collects from Tables of Observations continued for 82 years, that is from A.D. 1629 to 1711, that the number of Births in *London* was, at a *medium*, about 14,000 yearly : and likewise, that the number of *Males* to that of *Females*, or the facility of their production, is nearly as 18 to 17. But he thinks it the greatest weakness to draw any Argument from this against the Influence of *Chance* in the production of the two Sexes. For, says he, " Let 14,000 Dice, each having 35 faces, 18 white and 17 black, be thrown up, and it is great Odds that the numbers of white and black faces shall come as near, or nearer, to each other, as the numbers of Boys and Girls do in the Tables."

To which the short answer is this : Dr. *Arbuthnot* never said, " That supposing the facility of the production of a Male to that of the production of a female to be already *fixt* to nearly the Ratio of equality, or to that of 18 to 17 ; he was *amazed* that the Ratio of the numbers of Males and Females born should, for many years, keep within such narrow bounds : " the only Proposition against which Mr. *Bernoulli's* reasoning has any force.

But he might have said, and we do still insist that " as, from the Observations, we can, with Mr. *Bernoulli*, infer the facilities of production of the two Sexes to be nearly in a Ratio of equality ; so from this Ratio once discovered and *manifestly serving to a wise purpose*, we conclude the Ratio itself, or if you will the *Form of the Die*, to be an Effect of *Intelligence and Design*." As if we were shewn a number of Dice, each with 18 white and 17 black faces, which is Mr. *Bernoulli*'s supposition, we should not doubt but that those Dice had been made by some Artist ; and that their form was not owing to *Chance*, but was adapted to the particular purpose he had in View.

Thus much was necessary to take off any impression that the authority of so great a name might make to the prejudice of our argument. Which, after all, being level to the lowest understanding, and falling in with the common sense of mankind, needed no formal Demonstration, but for the scholastic subtleties with which it may be perplexed : and for the abuse of certain words and phrases ; which sometimes are imagined to have a meaning merely because they are often uttered.

Chance, as we understand it, supposes the *Existence* of things, and their general known *Properties* : that a number of Dice, for instance, being thrown, each of them shall settle upon one or other of its Bases. After which, the *Probability* of an assigned Chance, that is of some particular disposition of the Dice, becomes as proper a subject of Investigation as any other quantity or Ratio can be.

But *Chance*, in atheistical writings or discourse, is a sound utterly insignificant : It imports no determination to any *mode of Existence* ; nor indeed to *Existence* itself, more than to *non-existence* ; it can neither be defined not understood : nor can any Proposition concerning it be either affirmed or denied, excepting this one, " That it is a mere word."

The like may be said of some other words in frequent use ; as *fate, necessity, nature*, a *course of nature* in contradistinction to the *Divine energy* : all which, as used on certain occasions, are mere sounds : and yet, by artful management, they serve to found spacious conclusions : which, however, as soon as the latent fallacy of the *Term* is detected, appear to be no less absurd in themselves, than they commonly are hurtful to society.

I shall only add, that this method of Reasoning may be usefully applied in some other very interesting Enquiries : if not to force the Assent of others by a strict Demonstration, at least to the Satisfaction of the Enquirer himself : and shall conclude this

Remark with a passage from the *Ars Conjectandi* of Mr. *James Bernoulli, Part* IV, *Cap.* 4, where that acute and judicious Writer thus introduceth his Solution of the Problem for *Assigning the Limits within which, by the repetition of Experiments, the Probability of an Event may approach indefinitely to a Probability given,* " Hoc igitur est illud Problema, etc." *This, says he, is the Problem which I am now to impart to the Publick, after having kept it by me for twenty years :* new *it is, and difficult ;* but *of such excellent use, that it gives a high value and dignity to every other Branch of this Doctrine.* Yet there are Writers, of a Class indeed very different from that of *James Bernoulli,* who insinuate as if the *Doctrine of Probabilities* could have no place in any serious Enquiry : and that Studies of this kind, trivial and easy as they be, rather disqualify a man for reasoning on every other subject. Let the Reader chuse.

Index

Abraham the Maronite, 215
Abrégé des Combinations, 81
Académie des Sciences, Paris, 75, 120, 134
addition, 182 ; of fractions, 187–8
Alais, Prince Louis de Valois, Comte d', 196, 216, 223
Alcala de Henares, University of, 200
Alcuin of York, 29–30
Aldobrandini, Cardinal, 210
Aleae geometria, 97
algebraic signs, 48 *note*, 61
Allatio, Leon, 207, 211
alligation, rule of, 190
alphabetic divination, 18
Amama, Sixtin, 204, 225
Ambosius, Marius, 198
Annonciade and S. François de Paule, Convent of l', 74, 199, 200, 204, 226
Annuities on Lives, de Moivre's, 172
Apes urbanae, 207
Apianus, P., 61
Apollonius's conic sections, 202
Arab mathematics, 28
Arbuthnot, J., 116, 265
Archimedes, works of (Mersenne), 202
arithmetic, origins of, 29 *et seq.* ; Paccioli on, 37 ; Buckley on, 179 *et seq.*
arithmetic triangle, 34, 61, 81 *et seq.*, 94–5, 124, 146, 150, 246
Arithmetica integra, 61
Ars Conjectandi, 115, 118, 133 *et seq.*, 151, 153, 157, 267
articulate, 182
Artis magnae liber, 50
association, rule of, 189
astragalus, 2 *et seq.* ; scoring with, 7 *et seq.* ; divination by, 15–16 ; number used, 18–19
astrology, 18, 211
Aubrey's *Brief Lives*, 100, 103
Augustine, St., 26

Augustus, addiction to gaming, 8
Auvry, Father Jean, 204

backgammon, game of, 5, 18, 149
balla, Paccioli's game of, 37
Barberini, Cardinal Antoine, 210, 211
Barberini, Cardinal François, 64, 74, 210, 212, 223
Baronio, Cardinal Cesar, 227
Barrillon, Jean Jaques, 217
Basairdy, Father Pierre, 197
bassette, game of, 121, 144
Bauldri, Dom Michel, 214
Bede, the Venerable, 2, 29
Bell, E. T., 99, 109
Bentivoglio, Cardinal, 210
Beringhen, Henry de, 216
Bernoulli, James, 20, 115, 118–119, 121, 125, 130 *et seq.*, 140–3, 150–1, 153–5, 159, 161, 165, 171, 173, 176, 178, 254, 267 ; argument on probability, 136 *et seq.*
Bernoulli, John, 72, 121, 131, 133–4, 151, 155–7, 164–5, 171, 173
Bernoulli, Nicholas, the elder, 130
Bernoulli, Nicholas, the younger, 121, 140–1, 143, 151, 155, 157–9, 165, 170–1, 254, 265–7 ; preface to *Ars Conjectandi*, 133 *et seq.*
Bernoulli numbers, 135, 142 *et seq.*
Bernoulli's theorem, 136–7
Berovicius, Jean, 206
Bills of Mortality, 98 *et seq.*
Bills of Mortality, Natural and Political Observations on the, 100
binomial, sum of terms of, 168–9, 254 *et seq.*
biquadratic, solution of, 49
Bisci, Cardinal, 210
Blanchar, René, 197
board games, 4 *et seq.*

Boccaccio, Giovanni, 36
Boethius, 33, 50, 207
Bolduc, Rev. Father, 206, 214
Borghese, Cardinal, 210
Bouchard, Jean Jaques, 210
Boulenger, 220
Boulliau, Ismael, 210, 211, 215
Bouqueval, Marcel de, 218
Bourbon, Armand de, Prince de Conty, 215, 220
Bourbon, Henry de, Marquis de Verneuil, 222
Bourbon, Louis de, Comte de Soissons, 220
Bourdelot, Jean, 210, 221
Breda, Academy of, 208
Brief Lives, Aubrey's, 100
Browne, W., 116
Bruno, Father Jean, 200
Bruno, St., 212
Buckley, William, 2, 62, 123–4, 179 *et seq.*

Caesars, Lives of the, 8
Calcagnini, Celio, 55, 56, 59
calculus, differential, 140, 142, 171–2
Cambrai, Chronicle of, 14
Campanella, Father Thomas, 214
Candale, François de Foix de, 202
Carcavi, Pierre de, 74–5, 80, 82, 89–90, 111–13
card games, origin of, 20 ; Montmort on, 143–4
Cardano, Facio, 40 *et seq.*
Cardano, Geronimo, 8, 36, 38, 40 *et seq.*, 211
Chaillou, Father Olivier, 199
Chance of the Dyse, 20
Chanut, 223
Charlemagne, 30
Charles II, King of England, 102
Charles IX, King of France, 216, 223
Chastelier, Father, 198
Chaucer, *Pardoner's Tale*, 30
Cheyne, George, 164
Christ, horoscope of, 53, 56
Christian Church, view of sortilege, 14, 19 ; attitude to idea of randomness, 26 ; opposition to gaming, 30
Chrysippus, 23
Chu Shih-Chieh, 34
Cicero, 24–5, 26
Claudius, addiction to gaming, 9

Colla, Giovanni, 48–9
Colletet, François, 222
Colletet, Guillaume, 208, 222
combinatorial method, 24, 58, 62, 81 *et seq.*, 189 ; Wallis, 123 ; Montmort, 146 ; Galileo, 192 *et seq.* ; Pascal-Fermat-Carcavi letters, 233 *et seq.*
— notation, 147, 156
Commendius, 203
Commercium epistolicum, 159
concatenation, 185
conditional probability, 146, 166, 170
Connington, Susan, 39
Considerazione sopra il Giuoco dei Dadi, 64
Conty, Armand de Bourbon, Prince de, 215, 220
Coste, Hilarion de, 196
Cotes, R., 176, 259
Cours mathématique, 82
Craig, C., 154–5
Cramoisy, Gabriel and Sebastian, 196
Criton, George, 198
Cromwell, Thomas, 99
cubes, 236
cubic equations, solution of, 47 *et seq.*, 72
cypher, 181, 182, 188

Dante, *Purgatorio*, 35, 149
De Arte conjectandi in jure, 151
decimal scale, origin of, 1
De Divina Proportione (Paccioli), 36
De Divinatione, 24 *et seq.*
Delamotte, 14
De Malo Recentiorium Medicorum Medendi Usu, 45
De Mensura Sortis, 151, 156, 158, 165
denominator, 186–7
De Ratiociniis in Aleae Ludo, 113–5, 132, 161
Desargues, R., 74, 76, 220
Descartes, René, 74–8, 82, 111, 123, 131, 206, 211, 221, 225
De Subtilitate Rerum, 51, 57
detached coefficients, method of, 171
De Vetula, 32 *et seq.*, 58
De Vita Propria Liber, 42, 54
dice, early versions, 9 *et seq.*, 17, 18, 22
dice-games, Bishop Wibold on, 31–2 ; Huygens on, 112 *et seq.* ; Galileo on, 192–5 ; *see* points, problem of
Die Coss, 61

die-throwing, theoretical concepts derived from, 58 *et seq.*
Dierkens, 120
difference quotients of zero, 170, 171
Digby, Sir Kenelm, 217, 224
digit, 182, 185
Divina Commedia, 35
divination, 13 *et seq.*; alphabetical, 18 *note*
division, 182, 184, 185; of fractions, 187–8
Doctrine of Chances, 115, 138, 152, 161 *et seq.*, 254–67
dog, dice-throw, 8
Domesday Book, 98
Doni, Jean-Baptiste, 210, 211, 224
Du Bellay, Cardinal, 196
Du Chesne, Michel, 215
Du Chesne de Forest, Nicholas, 207, 216
Du Puy, Chevalier Cassian (Le Cau del Pozzo), 211, 218, 224
duration of play, 171, 173
Du Rel, Father Jean, 205, 226
Du Ryer, Pierre, 222
Du Val, Dr. André, 198
Dyse, Chance of the (fortune-telling poem), 20

Edward VI, King, 52, 180
Egyptian die, 18; tomb painting, 5
Elzevirs, Abraham and Bonaventure, 206
error, rule of, 190
Essai d'Analyse sur les Jeux de Hasard, 115, 140 *et seq.*, 165
Essai pour les Coniques, 76
Euclid, 76, 161, 202
Euripides, throw of, 8
Exercitationes mathematicae, 112, 132
exponential limit, 168
— series, 146

Fabry de Peiresc, Nicolas-Claude, 73, 209–10, 213, 218
fate, James Bernoulli on, 137
Fermat, Pierre de, 70 *et seq.*, 111, 113, 119, 153–4, 218, 224; problem of points, 83 *et seq.*, 229 *et seq.*; theory of numbers, 249
Ferrari, Ludovico, 45 *et seq.*
Ferreo, Scipio, 48, 72
Fibonacci, Leonardo, 31
figurate number, 156

figures, Hindu, 27; fourteenth-century German, 33; Buckley on, 181
Fiore, Antonio Maria, 48, 72
Fisher, R. A., 167 *note*
Fludd, Robert, 204
fluxions, calculus of, 171–2
Follengius, Jean Baptiste, 206
Fontenelle, 133
Fortuna, 21, 24, 144
Fouchy, Grandjean de, 163, 178
Fournier, Father George, 205
Fournival, Richard de, 33
fractions, 184, 186–8
Francis of Paula, St., 212
François de Sales, Mgr. (St.), 215
François, Father Jean, 214, 224, 226
Frazer, Sir James, 15
Frenicle, 81, 222
Frey, Jean Cecile, 212
Frizon's *Gallia Purpurata*, 213

Gagnières, Aimé de, 81, 82
Galileo Galilei, 61 *et seq.*, 69, 74, 149, 192 *et seq.*, 201, 224
Gallia Purpurata, Frizon, 213
Gama, Don Vasco Luis de, 216
Gamaches, Dr. Philippe de, 198
gambler's ruin problem, 119, 135, 170
gaming, origin of, 6
Gassendi, Pierre, 74, 205, 208–9, 211, 213
Generale Trattato, 38, 61–2
geometry, Mersenne's works on, 202; Pascal on, 97, 237, 252
Girard, 220
golden rule, 189–90
— theorem, James Bernoulli's, 136, 138, 140, 157, 174
Gondy, Mgr. Jean François de, 201
Gondy, Mgr. Henry de, 200
Got, Marquis de Rouillac de, 216
Gould, S. H., 54
Graunt, John, 100 *et seq.*, 120
Graves, Robert, 9 *note*, 18 *note*, 26
Greenwood, M., 100 *note*, 109
Grotius, Hugo, 221
Grounde of Artes, 61
Guériteau, Father Nicolas, 199

Habert de Montmor et de la Brosse, Henry Louis, 217
Halé, 218

Halley, E., 162, 171
halma, 4
Hamilton, John, 52
Harcourt Brown, E., , 74, 80
Harmonicon celeste, 211
Harmonie Universelle, 81, 201–2, 211
Hastings, J., 26
hazard, game of, 18, 34–5, 148–9, 195 *note*
Hébert, Father Pierre, 199, 200
heel-bone, 2 *et seq.*
Henry IV of France, 198
Hérigone, Pierre, 82
Herodotus, 5
Hévélius, Jean, 206, 223
Hindu arithmetic, 27
History of the Great Plague, 99
Hobbes, Thomas, 207, 225
Hoeffer, S., 69, 80, 139
Homer, 1, 4, 197
Hontan, Baron, 149
hope, game of, 149
hounds and jackals, game of, 4
hucklebone, 2 *et seq.*
Hudde, John, 119, 120, 146
Huygens, Christianus, 20, 85, 92, 110 *et seq.*, 125, 132 *et seq.*, 146, 150, 152, 154, 159, 161, 165
Huygens, Constantin, 223
Huygens, Ludovick, 111, 120

icosahedral die, 11
Ignatius, Loyola, St., 212
integer, 186–8
inverse probability, 133, 137–8

Jansenism, 77 *et seq.*

Kendall, M. G., 31, 38–9, 58
Khayyam, Omar, 34
King's College, Cambridge, 180
Kingsley, C., 68 *note*
knucklebone, 2 *et seq.*

La Cueva, Cardinal de, 210
La Flèche, college of, 74, 198, 224
Lagrange, 172
La Mote le Vayer, Felix, 196
La Mothe le Vayer, François de, 208, 221
Langrenus, Michel Florent, 206

La Nouë, Father François de, 205, 226
lansquenet, game of, 144
La Tour, Father de, 198
Laubespine, Charles de, 217
Laubespine, Claude de, 217
La Vérité des Sciences, 81, 201
Le Cau del Pozzo, 211, 218, 224
Le Fèvre, Nicolas, 227
Léger, St., 14
Leibnitz, Gottfried Wilhelm, 85, 92, 131–3, 140, 151, 153–4, 159, 164, 171–2
Leonardo da Vinci, 36, 37, 40 *et seq.*
Leonardo the Pisan, 31, 37, 47
Leyden University, 111, 206
Liber Abaci, 31, 37
Liber de Ludo Aleae, 43, 54 *et seq.*
Libri, 69
Lichfield, 180
Lobkowitz, Abbot Dom Jean Caramuel, 206
logarithmic spiral, 138
— series, 255 *et seq.*
Longueterre, de, 215
Louis XIII, 212, 214, 221, 227
Louis XIV, 120, 216, 221
Loyola, St. Ignatius, 212
Lucian, 17
Ludovico Sforza il Moro, 36
Lydia, 5

Maignan, Father Emanuel, 226
Magni, Rev. Father Valérien, 207, 223
Malebranche, Father Nicholas de, 141, 151
matching distribution, 145–6
Marrier, Dom Martin, 214
Marsile, Theodore, 198
Martin, Father Simon, 206
Mauduit, Jaques and Louis, 222
Mécaniques de Galilée, Les (Mersenne), 201
Megerlin, Peter, 131
Mendoza, Pierre Hurtado de, 206
Méré, Antoine Gombaud, chevalier de, 81 *note*, 84–5, 88–9, 125, 231, 235
Mersenne Academy, 73 *et seq.*, 91, 95
Mersenne, Julien, 197
Mersenne, Marie, 197
Mersenne, Marin, 73 *et seq.*, 81, 95 ; *Life of*, 196–228 ; list of works of, 201 *et seq.* ; social circle, 212 *et seq.* ; extent of correspondence, 223 *et seq.*

Mersenne, Pierre, 219
Meusnier, Pierre, 207
Micheria, Chiara, 41
Milton, John, 63, 64
Miscellanea Analytica, 172–4, 254
Mohammed ben Musa, 28, 51
Moivre, Abraham de, 115, 119, 121, 125, 134; 141, 151,159,161 *et seq.*, 254 *et seq.*; argument in probability, 138; friendship with Newton, 163; binomial exponential limit, 168 *et seq.*; contribution to theory, 172–3; approximation to sum of terms of binomial, 174 *et seq.*
Montmor et de la Brosse, Henry Louis Habert, Seigneur de, 217
Montmort, Pierre Rémond de, 115, 119, 133–4, 165–7, 170–1, 173; on theory of probability, 141 *et seq.*
Montoya, Father Luc de, 206
Montucla, 51, 69
Morin, Jean Baptiste, 74, 210, 219
Morley, Henry, 41, 54
Mortality, Bills of, 98 *et seq.*
Moulière, Jeanne, 197
multinomial theorem, 170
multiplicand, 183
multiplication of fractions, 187–8
— tables, Paccioli's, 37
Munos and Spinossa, Thomas, 196, 205, 226
Mydorge, Claude, 202, 219
Mylon, P., 111–13

Natural and Political Observations on the Bills of Mortality, 100
Naudé, Gabriel, 207, 211, 221
Navarre, Royal College of, 213, 215
Newbold, E. M., 178
Newton, Sir Isaac, 35, 123 *et seq.*, 140, 151, 159, 162, 169, 176, 259; and Pepys, 125 *et seq.*; and de Moivre, 163 *et seq.*
Niceron, Father Jean François, 203, 219, 226
Nightingale, Florence, 103
Niphus, Augustin, 211
Nivernois, Duc de, 200
Noel, Father Estienne, 207, 214
normal curve, 174 *et seq.*
noyaux, jeu des, 149
Nozzolini, 67

numbers, origins of, 1–2, 27 *et seq.*; ancient Greek system, 12 *note*; Buckley on, 181; compound, 182; integer, 186; sum of natural, 236; theory of (Fermat), 249
numerator, 186–7
nuts, game of, 149

odds and evens, divination, 15
Odyssey, 1
ombre, game of, 144, 146
Ore, O., 54, 58
Ory, Sanson, 197
osselot, 9, 22
Oxford, University of, 123, 204
Ozanam, Jacques, 125 *note*, 161

Paccioli, Fra Luca di Borgo, 36, 38, 47, 48, 50
Padua, University of, 43–4
Pallote, Cardinal, 210
Pamphilio, Cardinal, 210
Pappus, 202
Paris, University of, 196, 198, 212, 214, 215, 219
particles, 186
Pascal, Blaise, 72 *et seq.*, 111, 113, 153–4, 159, 161 *note*, 220; problem of points, 83 *et seq.*; correspondence with Fermat, 229 *et seq.*
Pascal, Etienne, 75 *et seq.*, 220
Patroclus, 4
Pavia, University of, 40, 43
Pearson, Karl, 174 *note*, 178
Peckham, John, 41
Peiresc, Nicolas Claude Fabri de, 73, 209–10, 213, 218
Pell (Pellius), John, 111, 208
Pelletier, Jacques, 208
Pepys, Samuel, 102, 125 *et seq.*
Périer, M., 76, 78
Périer, Mme Gilberte, 76
Petit, 207, 222
Petrarch, 40
Peverone, G. F., 38
pharaoh, game of, 142, 144
Phelipeaux, Father Jean, 198, 205, 214
Philosophiae Christianae Principia Mathematicae, 154
Picot, 221
picquet, game of, 144, 146

Pisa, University of, 63–4
Place Royale, Paris, Convent of St. François de Paule, near the, 74, 199, 200, 204, 226
plague deaths in London, 105
Plato, 137
playing cards, 20
Plutarch, 23 *note*, 197
point, 181, 185
points, problem of, 37–9, 79, 88, 157; origin of, 28; Fermat-Pascal-Carcavi letters, 83 *et seq.*, 229 *et seq.*; Huygens and, 113 *et seq.*; de Moivre and, 167–8
Poisson, Pierre, 223
Poisson, Siméon Denis, 168
Poisson's binomial exponential limit, 168
population of London, 107–8
Port-Royal, Abbey of, 77–9, 82
Practica Arithmetica et Mensurandi Singularis, 47
Prieur, Father Jean, 200
primes, 95, 187, 247, 249
Principia Mathematica (Newton's), 162
probability, early reasoning on, 24 *et seq.*; theory of, derived from dice-throwing, 35, 59, 62; calculus of: Fermat and, 71 *et seq.*; Pascal and, 75 *et seq.*; James Bernoulli and, 132 *et seq.*; Montmort and, 142 *et seq.*; *see also* points, problem of
probability, conditional, 146, 168, 170
probability set, fundamental, enumeration of: Peverone, 39; Cardano/Ferrari, 58; Graunt, 106
product, 183
progression, arithmetical, 184; geometrical, 185
Propertius, 17
proportion, 185
Ptolemy, 6, 18, 50
Pythagoras, 19

Quakers, 99
quaternary marks, 181
quinquenove, game of, 120, 148
quotient, 184, 185, 190

randomisation, 12, 13, 21 *et seq.*, 24
Rangueil, Father Claude, 206
rebirth chart, 17
Recherche de la Vérité, 141
Record, Robert, 61
reduction, 187

Regnault, Father Robert, 226
remainder, 182, 184–7, 190
Reynault, Father Theophile, 212, 224
Rigaud, Nicolas, 218
Roannes, Duc de, 252
Robartes, Francis, 151, 166
Robert, Claude, 205, 213
Roberval, Gilles Personne de, 74–5, 82, 85, 91–2, 94, 95, 111–12, 219, 231, 240, 247, 248
Royal Society of London, 75, 102, 119, 121, 151, 153, 159, 162–3
Rudolff, Christoff, 61
rule of three, 189

Saint Charles, Father Louis Jacob de, 206
Sainte-Croix, 95
Sainte Marthe, Louis de, 205, 221
Sainte Marthe, Scévole de, 205, 221
Saint Michel, Chevalier de, 221–2
Saint Romvald, Dom Pierre de, 205
Sales, Mgr. (St.) François de, 215
Sambursky, S., 23 *note*, 26
Sarton, G., 34, 39
Saturnalia, 7
Saveur, Joseph, 121
Sceptics, 201, 209
Scheinerus, Christofle, 206
Schooten, Francis van, 111–14, 124, 132, 206, 225
Scott, J. F., 47–8 *notes*, 54, 132, 139
Selden, John, 207, 225
Selenus, 202
sex ratio, births in London, de Moivre on, 265
Simpson, Thomas, 172
Smith, D. E., 39, 82 *note*
snakes and ladders, game of, 5
snap, game of, 145
Sopra le Scoperte dei Dadi, 64, 192 *et seq.*
Sorbonne, 78, 79, 198, 213, 214
sortilege, 13 *et seq.*
square roots, 185–6, 188
Stampisen, 111
statistical method, origins of, in England, 98 *et seq.*
Stella, Jean Tileman, 220
Steven, Simon, 81
Stifel, Michael, 61, 81
Stirling, James, 172, 175–6, 256, 259
Stirling's theorem, 173 *et seq.*
Storer, Jean, 54

Struyck, Nicholas, 115, 119, 121
subtraction, 182 *et seq.*
Suetonius, 8–9, 17
*Summa de Arithmetica, Geometria, Proportioni
 e Proportionalità* (Paccioli), 37
Sylvestro, Rudolfo, 53

talus, 2, 17–18
Tartaglia, Nicolas Fontana, 38, 40 *et seq.*,
 47 *et seq.*, 62
Taylor, Sherwood, 69
tessera, 6 *note*
Theodosius, spherics of, 161, 202
throwing sticks, 11
Thurnisius, brothers, 133
Tillotson, Dean J., 104–5
Todhunter, I., 116, 155
Tollet, George, 129
Torricelli, Evangelista, 74, 224
Tournere, Sieur, 219
treize, game of, 144
Trojan Wars, 6
Trouillard, René, 223
Turkey, excavations in, 5

vacuum, 203, 207
Valençay, Marquis d' Estampes, 217
Valerius, Luc., 203

Vallombrosa, 63, 64
Valois, Louis Emmanuel de, 196, 216,
 223
Venus throw, 8, 25
Viète, François, 202, 206, 211
Vinci, Leonardo da, 36, 37, 40 *et seq.*
Voisin, Joseph, 215
Voltaire, 79
Vulson, Marc de, 221

Waddell, Helen, 29 *note*
Waddell, L. A., 16
Walker, H. M., 178
Wallis, J., 123 *et seq.*
Waters, W. G., 44, 54
Wesley, John, 14
Whetstone of Witte, 61
Wibold, Bishop, 31
Wollenschläger, K., 178

Xenocrates, 23 *note*, 81

Ysambert, Dr. Nicolas, 198

zara, game of, 195 *note*
zero, differences of, 156